Peter Christian Bruns

Multi-scale chiral dynamics

Peter Christian Bruns

Multi-scale chiral dynamics

Some new developments in effective quantum field theories

Südwestdeutscher Verlag für Hochschulschriften

Impressum / Imprint

Bibliografische Information der Deutschen Nationalbibliothek: Die Deutsche Nationalbibliothek verzeichnet diese Publikation in der Deutschen Nationalbibliografie; detaillierte bibliografische Daten sind im Internet über http://dnb.d-nb.de abrufbar.

Alle in diesem Buch genannten Marken und Produktnamen unterliegen warenzeichen-, marken- oder patentrechtlichem Schutz bzw. sind Warenzeichen oder eingetragene Warenzeichen der jeweiligen Inhaber. Die Wiedergabe von Marken, Produktnamen, Gebrauchsnamen, Handelsnamen, Warenbezeichnungen u.s.w. in diesem Werk berechtigt auch ohne besondere Kennzeichnung nicht zu der Annahme, dass solche Namen im Sinne der Warenzeichen- und Markenschutzgesetzgebung als frei zu betrachten wären und daher von jedermann benutzt werden dürften.

Bibliographic information published by the Deutsche Nationalbibliothek: The Deutsche Nationalbibliothek lists this publication in the Deutsche Nationalbibliografie; detailed bibliographic data are available in the Internet at http://dnb.d-nb.de.

Any brand names and product names mentioned in this book are subject to trademark, brand or patent protection and are trademarks or registered trademarks of their respective holders. The use of brand names, product names, common names, trade names, product descriptions etc. even without a particular marking in this work is in no way to be construed to mean that such names may be regarded as unrestricted in respect of trademark and brand protection legislation and could thus be used by anyone.

Verlag / Publisher:
Südwestdeutscher Verlag für Hochschulschriften
ist ein Imprint der / is a trademark of
OmniScriptum GmbH & Co. KG
Heinrich-Böcking-Str. 6-8, 66121 Saarbrücken, Deutschland / Germany
Email: info@svh-verlag.de

Herstellung: siehe letzte Seite /
Printed at: see last page
ISBN: 978-3-8381-1600-6

Zugl. / Approved by: Bonn, Universität Bonn, Diss., 2008

Copyright © 2010 OmniScriptum GmbH & Co. KG
Alle Rechte vorbehalten. / All rights reserved. Saarbrücken 2010

"Natur ist, vor und nach der Quantentheorie, das mathematisch zu Erfassende; selbst was nicht eingeht, Unauflöslichkeit und Irrationalität, wird von mathematischen Theoremen umstellt. In der vorwegnehmenden Identifikation der zu Ende gedachten mathematisierten Welt mit der Wahrheit meint Aufklärung vor der Rückkehr des Mythischen sicher zu sein."

Th. W. Adorno, M. Horkheimer: Dialektik der Aufklärung [1].

Multi-scale chiral dynamics

In this work, we examine two possible extensions of the low-energy effective field theory of quantum chromodynamics (QCD), called chiral perturbation theory (ChPT). Both those extensions are related to the presence of resonances in the effective field theory, which in its original formulation contains only the (pseudo-) Goldstone bosons of the spontaneously broken chiral symmetry of QCD (and, in a further formulation, also the baryon ground-state octet) as its degrees of freedom. The additional mass scales (the masses of the resonances) generally lead to complications concerning the structure of the effective theory. In particular, the power counting scheme, which determines the ordering of the perturbation series, can only be maintained using specific renormalization prescriptions for loop graph contributions. Such renormalization schemes will be worked out in chapters 2 and 3. In chapters 4 and 5, the resonance fields will not be included explicitly in the theory. Instead, we will employ a nonperturbative method (the Bethe-Salpeter integral equation) to describe the meson-baryon scattering amplitude in a resonance-dominated process (kaon electroproduction). Unlike perturbation theory, this nonperturbative ansatz is suited to generate resonances dynamically. However, the price to be paid is that some field-theoretical constraints are usually not met in such nonperturbative approaches. It is the aim of the analysis in chapter 4 to devise a method which implements the principle of gauge invariance in this extension of ChPT. In chapter 5, we will apply our method to the description of kaon electroproduction on the proton, a process under experimental investigation at several facilities at e.g. MAMI, ELSA, Spring-8 and GRAAL.

Contents

1 Introduction   1
   1.1 Quantum chromodynamics 1
   1.2 Chiral perturbation theory 5
   1.3 Baryon chiral perturbation theory 12
   1.4 Inclusion of resonance fields 15
   1.5 Unitarized chiral perturbation theory 17

2 Infrared regularization with vector mesons and baryons   23
   2.1 Introduction 23
   2.2 IR regularization in the pion-nucleon system 24
   2.3 IR regularization for vector mesons and pions 26
   2.4 Pion-nucleon system with explicit meson resonances 30
   2.5 Application: Axial form factor of the nucleon 34
   2.6 Summary 40

3 Quark mass dependence of the mass of the Roper(1440)   42
   3.1 Introduction 42
   3.2 Effective Lagrangian 43
   3.3 Chiral corrections to the Roper mass 44
   3.4 Quark mass dependence of the Roper mass 48

4 Gauge invariance in unitarized ChPT   51
   4.1 Introduction 51
   4.2 Complex scalar fields 52
   4.3 Weinberg-Tomozawa term 55
   4.4 Higher order interaction kernels 59
   4.5 Unitarity 60
   4.6 Conclusions 63

5 Threshold kaon photo-and electroproduction   64
   5.1 Introduction 64
   5.2 Bethe-Salpeter equation 66
   5.3 Photo- and electroproduction 71
   5.4 Results 77
   5.5 Conclusions and outlook 84

6 Outlook   88

A	Explicit expression for I_{MV}^{IR}	90
B	Alternative derivation of I_{MBV}^{IR}	92
C	Decomposition of infrared regularized loop integrals	94
D	Ward identity for general interaction kernels	96
E	Solution of the Bethe-Salpeter equation	97
F	UV-divergences in the Bethe-Salpeter equation	99
G	Bethe-Salpeter approach with off-shell kernel	102
H	Loop integrals for the electroproduction amplitude	104
I	Decomposition of the electroproduction amplitude	109
J	Invariant amplitudes for the electroproduction process	114

Chapter 1

Introduction

1.1 Quantum chromodynamics

It is widely believed by theoretical physicists today that all fundamental interactions observed in Nature so far can be described by quantum field theories. In order to make this assumption plausible, we shall quote the following 'theorem' stated by S. Weinberg:

"It is very likely that any quantum theory that at sufficiently low energy and large distances looks Lorentz invariant and satisfies the cluster decomposition principle[1] will also at sufficiently low energy look like a quantum field theory." [2]

There is a set of quantum field theories, called the Standard Model, that describes very well the interactions of the fundamental constituents of matter, fermions called leptons and quarks, with the force fields pertaining to the electromagnetic, the weak and the strong interaction. By the time of this writing, there is no clear evidence for observations contradicting the Standard Model. However, there is little doubt among theoretical physicists that this model will have to be modified in one way or the other. Notably, gravity is not included in the Standard Model. In this work, we are mostly concerned with Quantum Chromodynamics (QCD), which is a non-Abelian gauge field theory. It is the part of the Standard Model that describes the strong interactions between quarks and gluons, the latter being the gauge field quanta of the theory. The QCD Lagrangian is given by

$$\mathcal{L}_{\text{QCD}} = \bar{q}^i_f \{i\gamma^\mu D_{\mu,ij} - \delta_{ij} m_f\} q^j_f - \frac{1}{4} G^a_{\mu\nu} G^{a,\mu\nu} - \frac{g^2\theta}{64\pi^2} G^a_{\mu\nu} \tilde{G}^{a,\mu\nu}. \qquad (1.1)$$

In the above expression, and in the following, a summation over repeated indices is always implied (unless stated otherwise). In eq. (1.1), D is the gauge-covariant derivative

$$D_{\mu,ij} = \delta_{ij} \partial_\mu + ig t^a_{ij} A^a_\mu.$$

The quark fields are collected in the spinor q. There are six different kinds of quarks, labeled by the 'flavor' index $f = \{u, d, c, s, t, b\}$. To each flavor, a different quark mass parameter

---
[1] Loosely speaking, this principle requires that "distant experiments give uncorrelated results" [2]. For a more rigorous statement, see chapter 4 of [3].

m_f is associated. Quarks of flavor u, c and t carry an electric charge of $+2/3$, while d, s, b-quarks have charge $-1/3$ (in units of the elementary charge $e = 1,602 \cdot 10^{-19} C$). Moreover, there is a quantum number named color assigned to the quarks, labeled by the color indices $\{i, j, \ldots\}$ which run over three colors r, b, g. It is this color charge which is the source of the strong interaction mediated by the gluon gauge fields A_μ^a. The strength of this interaction is measured by the coupling parameter g. The gauge group is $SU(3)_c$, the set of special unitary transformations acting in color space, and the corresponding Lie algebra is spanned by the generators $t^a = \lambda^a/2$, where λ^a denotes the well-known Gell-Mann matrices. By definition, the generators are subject to the algebra $[t^a, t^b] = if^{abc}t^c$, where f^{abc} are the $SU(3)_c$ structure constants. From the gluon fields one can build the gauge-covariant field strength tensor

$$G_{\mu\nu}^a = \partial_\mu A_\nu^a - \partial_\nu A_\mu^a - gf^{abc} A_\mu^b A_\nu^c$$

and its dual,

$$\tilde{G}^{a,\mu\nu} = \epsilon^{\mu\nu\alpha\beta} G_{\alpha\beta}^a \quad .$$

Due to the presence of the Levi-Civita tensor ϵ in the definition of the dual field strength tensor $\tilde{G}$, the last term in the QCD Lagrangian of eq. (1.1) would be responsible for CP violation in strong interaction reactions. As there is no experimental evidence for this, we will assume henceforth that the so-called vacuum angle θ equals zero.

In the following chapters, our interest will be focused on the analysis of QCD at low energies. There, the c, t and b quarks, being of considerably larger mass than the remaining three quark flavors, are frozen, i.e. their presence reflects itself only in the form of local couplings of the other fields of the theory, as a consequence of the Heisenberg uncertainty principle. Consequently, the heavy quarks can be integrated out of the theory (in the sense of the path integral), leaving us with only the light quark flavors u, d, s and the gluons as dynamical degrees of freedom. In many applications, even the s-quark is integrated out, and only the two lightest flavors u, d remain as explicit degrees of freedom of the theory.

What is more important is the fact that the coupling g is not a constant, but depends on the energy scale of the physical reaction under consideration. The reason for this is that g is subject to renormalization, due to gluon and quark loop contributions to QCD matrix elements. For energies of at least a few GeV, the coupling is small, so that a perturbative expansion of the required matrix elements in powers of g will generally be applicable. This situation is usually referred to as asymptotic freedom: for very high energies (or equivalently, at short distances), the quarks effectively behave like free particles [4, 5]. In the region of low energies (energies less than about a GeV), however, the coupling is large, which renders a direct perturbative treatment of the theory useless. One direct way to proceed is given by the method of Lattice QCD, which is the attempt to evaluate the QCD path integral numerically on a discrete space-time lattice of finite volume, employing Monte Carlo sampling. Comprehensive introductions to this field can be found in [6, 7]. At present, most lattice simulations of QCD are restricted to pion masses $M_\pi \geq 250$ MeV (whereas in the real world $M_\pi \sim 140$ MeV), due to the growth of computational cost with increasing lattice size. One therefore needs a systematic way to extrapolate the lattice results to physical masses. We will provide an example for such an extrapolation formula in chapter 3. In this work, we do not make use of lattice simulation techniques. Instead, we shall follow another path that leads to a model-independent description of QCD Green functions, which is provided by the framework of effective field theories.

The effective degrees of freedom at low energies are the hadrons (baryons and mesons) formed from the quarks and gluons. This observation, together with the theorem by Weinberg cited above, leads to the following idea. Assuming that QCD is the correct choice as the underlying theory of strong interactions, and knowing the effective degrees of freedom emerging from this underlying theory at low energies, we can try to construct the most general quantum field theory (with the hadrons as fundamental fields appearing in the corresponding Lagrangian) that is consistent with all the symmetries and symmetry-breaking patterns of $\mathcal{L}_{\text{QCD}}$. According to Weinberg's approach to quantum field theories (first spelled out in [8]), this is the most general ansatz for a quantum theory describing low-energy hadron interactions consistent with Lorentz invariance, the cluster decomposition principle, and the symmetries specific for QCD. The effective Lagrangian will contain infinitely many terms, which are classified as unrenormalizable in the usual sense. Therefore, one needs an ordering scheme that allows to select the terms which give the leading contributions to the low-energy expansion of QCD matrix elements (or, more generally, to the low-energy expansion of the generating functional Z). Details on the construction and ordering of the effective Lagrangian will be given in the next section. For the rest of this section, we will focus on an approximate symmetry of the QCD-Lagrangian, which will turn out to be of fundamental importance for the construction of the effective field theory. As mentioned above, it is legitimate to keep only the three lightest quark flavors as active degrees of freedom when considering QCD at low energies. The masses of the remaining quarks are known to be very small compared to the typical hadronic scale, taken as e.g. the proton mass of roughly 1 GeV. In a first step, one can thus approximate the QCD Lagrangian by

$$\mathcal{L}_0 = i\bar{q}\slashed{D}q - \frac{1}{4}G^a_{\mu\nu}G^{a,\mu\nu}, \tag{1.2}$$

and treat the mass terms of the light quarks as a perturbation. In the last equation, the quark spinor q now only contains the light quarks, $q = (u, d, s)^T$. Furthermore, we used the common abbreviation $\slashed{D} \equiv \gamma_\mu D^\mu$, and left out color and flavor indices. The quark spinor q can be decomposed into so-called left-handed and right-handed components, making use of the chirality projectors for Dirac particles: $q = q_L + q_R$, with

$$q_L = \frac{1}{2}(1 - \gamma_5)q, \qquad q_R = \frac{1}{2}(1 + \gamma_5)q.$$

In the full QCD-Lagrangian, left-and right-handed quark fields are coupled by the mass terms. In $\mathcal{L}_0$, however, the components of different chirality do not communicate - they appear in a completely decoupled form: Standard Dirac matrix algebra yields

$$i\bar{q}\slashed{D}q = i\bar{q}_L\slashed{D}q_L + i\bar{q}_R\slashed{D}q_R.$$

As a consequence, $\mathcal{L}_0$ is invariant under independent global unitary transformations of the left-and right-handed quark fields in flavor space,

$$q_L \to L q_L, \qquad q_R \to R q_R,$$

where $L \in U(3)_L, R \in U(3)_R$. The full symmetry group can be decomposed according to

$$U(3)_L \times U(3)_R = SU(3)_V \times SU(3)_A \times U(1)_V \times U(1)_A.$$

The subscripts V and A stand for vector and axial-vector transformations, defined by $R = L$ and $R = L^\dagger$, respectively. The symmetry associated with this transformation group is usually referred to as 'chiral symmetry', and the limit of the QCD Lagrangian in which $m_u, m_d, m_s \to 0$ is called the 'chiral limit' of QCD.

Given that the strategy to treat the light quark mass term as a small perturbation is valid, one must now ask whether the above symmetry of the Lagrangian is (at least approximately) realized in Nature, so that e.g. the hadron spectrum can be organized in multiplets corresponding to the symmetry group. The irreducible subgroup $U(1)_V$ can in fact be associated with baryon number conservation (note that this particular subgroup is not only a symmetry for the Lagrangian in the chiral limit, but also for nonvanishing quark masses). In contrast to that, the axial $U(1)_A$ symmetry is broken by an anomaly [9, 10]. This means that these transformations, though leaving the classical action invariant, do not form a symmetry of the quantum theory, because they induce a change in the measure of the path-integral [11].

Let us turn to the remaining symmetry group $SU(3)_L \times SU(3)_R = SU(3)_V \times SU(3)_A$. As explained in [12], there are very convincing reasons to believe that the vectorial subgroup $SU(3)_V$ is left unbroken in the chiral limit. Indeed, the observed lowest-lying hadron states form approximate multiplets of this flavor symmetry. Small violations of this symmetry pattern can be traced back to light quark mass differences, and to effects due to the electromagnetic interaction. There are, on the other hand, strong indications that the $SU(3)_L \times SU(3)_R$ symmetry is spontaneously broken down to its vectorial subgroup $SU(3)_V$. It can readily be seen from the hadron spectrum that the full $SU(3)_L \times SU(3)_R$ symmetry can not be realized (not even approximately), as this would imply a parity-doubling of the hadron-states, which is not observed. This leads us to conjecture that the symmetry in question is not realized in the familiar Wigner-Weyl mode, leading to symmetry multiplets in the energy spectrum, but in the Nambu-Goldstone mode [13, 14], in which case the symmetry of the Lagrangian is not shared by the ground state of the theory. This spontaneous breaking of a symmetry necessarily leads to the existence of massless spin zero particles, called Goldstone bosons. More precisely, the number of different Goldstone boson fields present in the theory is given by the number of generators of the spontaneously broken symmetry group, and the Goldstone bosons have the quantum numbers of those generators. This is the content of Goldstone's theorem [14]. The number of generators of $SU(N)$ is $N^2 - 1$, consequently, in our case of the product group $SU(3)_L \times SU(3)_R$ broken down to $SU(3)_V$, there would have to be eight massless Goldstone bosons with the quantum numbers of the generators of $SU(3)_A$, i.e. an octet of eight pseudoscalar mesons. They would have to be massless in the idealized case where the light quark masses are exactly zero. For nonvanishing, but small, quark masses, they should still be lighter than the other hadrons, the masses of which is very roughly of the order of the typical hadronic scale referred to above, that is, in the GeV range (were this not the case, the treatment of the light quark masses as a small perturbation would be questionable). Indeed, the lightest hadrons one finds are the pseudoscalars $\pi^\pm, \pi^0, K^\pm, K^0, \bar{K}^0, \eta$ with masses given by $M_\pi \approx 138$ MeV, $M_K \approx 495$ MeV and $M_\eta \approx 547$ MeV [15]. This has to be compared to the mass of the lowest-lying hadron that is not an approximate Goldstone boson, the rho resonance, with mass $M_\rho \approx 770$ MeV. We can expect that correction terms with respect to the chiral limit are suppressed by factors of roughly $M_K/M_\rho \approx 0.65$ for chiral $SU(3)_L \times SU(3)_R$ symmetry, while the suppression factors are of order $M_\pi/M_\rho \approx 0.18$ when considering only chiral $SU(2)_L \times SU(2)_R$ broken down to $SU(2)_V$. In the latter case, the s-quark is integrated out, and there is only a triplet of Goldstone bosons, namely, the pions. Obviously, it is the $SU(2)$-case where we can expect a better

convergence of the perturbative expansion around the chiral limit[2].
In summary, we are dealing here with a spontaneously broken approximate symmetry [16], with explicit symmetry breaking given by nonvanishing quark masses. In the next section, we will show how this symmetry pattern is used to arrive at a systematic low-energy expansion of QCD matrix elements in terms of an effective field theory involving the lowest-lying hadron states.

1.2 Chiral perturbation theory

In this section, we will see how to construct the effective field theory of QCD which describes the interaction of the Goldstone bosons at low energies. To this end, it is very useful to promote the global $SU(3)_L \times SU(3)_R$-symmetry studied in the previous section to a local symmetry, by means of the technique of external sources. For a general introduction to functional methods in quantum field theory, in particular to the method of external sources, we refer to the textbook [17]. Let us introduce an additional term in $\mathcal{L}_0$, eq. (1.2), which contains some externally assigned sources v, a, s, p, being of vector, axial-vector, scalar and pseudoscalar character, respectively:

$$\mathcal{L}' = \mathcal{L}_0 + \bar{q}[\gamma_\mu(v^\mu + \gamma_5 a^\mu) - (s - ip\gamma_5)]q. \tag{1.3}$$

The external fields v, a, s, p are hermitean matrices in flavor space. We do not include the singlet components of v and a, so that $\text{tr } v^\mu = \text{tr } a^\mu = 0$. With this condition, the generating functional, defined by

$$e^{iZ[v,a,s,p]} = \langle 0 | T \exp\left(i \int d^4x \bar{q}[\gamma_\mu(v^\mu + \gamma_5 a^\mu) - (s - ip\gamma_5)]q\right) | 0 \rangle, \tag{1.4}$$

is invariant under local $SU(3)_V \times SU(3)_A = SU(3)_L \times SU(3)_R$ transformations, provided that the external source fields are transformed according to

$$v_\mu + a_\mu \equiv r_\mu \to R(v_\mu + a_\mu)R^\dagger + iR\partial_\mu R^\dagger, \tag{1.5}$$

$$v_\mu - a_\mu \equiv l_\mu \to L(v_\mu - a_\mu)L^\dagger + iL\partial_\mu L^\dagger, \tag{1.6}$$

$$s \to RsL^\dagger, \quad p \to RpL^\dagger, \tag{1.7}$$

where $L \in SU(3)_L$, $R \in SU(3)_R$. In eq. (1.4), T is the usual time ordering operator, and Z is the generating functional of connected Green functions (see e.g. [17]). The quark mass matrix, given by $\mathcal{M} = \text{diag}(m_u, m_d, m_s)$, will be included in the external scalar field s, so that $s = \mathcal{M} + \ldots$, and is thus treated as an external perturbation. The vacuum expectation value on the r.h.s. of eq. (1.4) is represented as a path integral over all quark and gluon field configurations. In a covariant quantization procedure, the measure of the path integral must also include the Fadeev-Popov determinant. For a detailed account on the quantization of non-abelian gauge field theories, we again refer to standard textbooks [16–18], and references cited therein.

The rationale for the introduction of external sources is twofold. First, every Green function of quark current operators can be derived from the generating functional of eq. (1.4) by applying

---

[2]Actually, it turns out that, in the pure mesonic sector of the effective theory, the corrections are suppressed by factors of $(M_K/M_\rho)^2 \approx 0.4$ and $(M_\pi/M_\rho)^2 \approx 0.03$, so that the above estimates are too pessimistic in that case.

multiple functional derivatives with respect to v, a, s and p to this functional. Secondly, the formalism of external sources is well suited for the implementation of additional interactions. For example, an electromagnetic coupling to the quark fields can be introduced via $v_\mu = -eQ\mathcal{A}_\mu$, where $\mathcal{A}$ is the photon field and $Q = \text{diag}(2/3, -1/3, -1/3)$ is the quark charge matrix.

The key idea of the effective field theory treatment of QCD can now be expressed by means of the formula

$$e^{iZ[v,a,s,p]} = \int [dU] \exp\left(i \int d^4 x \mathcal{L}_{\text{eff}}(U, v, a, s, p)\right). \tag{1.8}$$

The path integral measure is written as $[dU]$, where U comprises the Goldstone boson fields in a suitable way, as specified below. By expressing the generating functional of QCD in terms of a field theory involving only the Goldstone bosons as dynamical degrees of freedom, one makes use of the assumption that the low-energy structure of QCD is dominated by these particles. It is important to note that formula (1.8) is not suited for an analysis of QCD at energies where other degrees of freedom, like e.g. baryons or meson resonances like the rho, become important for the dynamics. In eq. (1.8), all those high-energy degrees of freedom have been integrated out, their effects being encoded in the local terms of $\mathcal{L}_{\text{eff}}$. Moreover, it is crucial here that QCD confines the quarks and gluons, so that these never appear as asymptotic states. Otherwise, massless gluons and light quarks could yield important contributions to the long-range sector of the strong interaction physics. Due to confinement, the long-range interaction is presumably dominated by the (approximately) massless Goldstone bosons. We have already seen in the previous section that the Goldstone bosons can be identified (in the $SU(3)$ case) with the π,K and η mesons.

It has been shown by Leutwyler [19] that, in the absence of anomalies, the Ward identities of QCD ensure that the low energy properties of the QCD Green functions can be analyzed in terms of an effective field theory employing an effective Lagrangian symmetric under local $SU(3)_L \times SU(3)_R$ transformations. Given that the Goldstone bosons are the only important degrees of freedom at very low energies, this justifies eq. (1.8), as well as the seemingly artificial construction involving external source fields and a local version of the $SU(3)_L \times SU(3)_R$ invariance, which is only a global symmetry for the original Lagrangian of QCD, eq. (1.1).

In the remainder of this section, we will see how to construct the effective Lagrangian $\mathcal{L}_{\text{eff}}$ which describes the interaction of the Goldstone bosons. As noted in the previous section, the effective Lagrangian must share all the symmetries of $\mathcal{L}_{\text{QCD}}$. In particular, the Goldstone boson fields ϕ should transform under some realization of the chiral symmetry group $G \equiv SU(3)_L \times SU(3)_R$, and the invariants formed from these fields are then candidates for terms in the effective Lagrangian (of course, these terms must also be consistent with Lorentz invariance and invariance under the discrete symmetry transformations C, P and T).

We start in a fairly general way and write the transformation of the Goldstone boson fields under the group G as

$$\phi \to \phi' = F(g, \phi), \quad g \in G, \tag{1.9}$$

where the function F is required to be continous, and subject to the group homomorphism property

$$F(g_1, F(g_2, \phi)) = F(g_1 g_2, \phi). \tag{1.10}$$

Furthermore F is defined such that $F(e, \phi) = \phi$, where e is the unit element of G. The vacuum is given by the state where no ϕ excitation is present, $\phi \equiv 0$. Consider the set H of elements of G that leave the vacuum invariant: $h \in H \Leftrightarrow F(h,0) = 0$. From the composition law of eq. (1.10) we note that H is a subgroup of G. We have assumed a scenario of spontaneous symmetry breaking where $SU(3)_A$ transformations do not leave the vacuum invariant, while $SU(3)_V$ transformations do. Therefore, the subgroup H must be identified with $SU(3)_V$. Every configuration ϕ can be reached by a group transformation acting on the vacuum, so that the function

$$g \to F(g,0) = F(gh,0) \quad \forall h \in H$$

can be seen as a mapping from the coset space G/H to the space of Goldstone boson fields. It is injective because

$$F(g_1,0) = F(g_2,0) \Rightarrow 0 = F(g_1^{-1}, F(g_1,0)) = F(g_1^{-1}, F(g_2,0)) = F(g_1^{-1}g_2, 0) \Rightarrow g_1^{-1}g_2 \in H.$$

Moreover, the dimension of the coset space G/H is equal to the dimension of the space of Goldstone boson fields, which is given by the number of generators of the transformation group which does not leave the vacuum invariant. Consequently, we have established a one-to-one correspondence of the Goldstone boson fields with the elements of G/H.

Every element of G can be written as

$$g = (g_L, g_R) = (1, g_R g_L^{-1})(g_L, g_L) \equiv qh,$$

which amounts to the choice of a representation $(1, g_R g_L^{-1})$ for $q \in G/H$. The action of the group G on G/H is then given by

$$(L, R)(1, g_R g_L^{-1}) = (L, R g_R g_L^{-1}) = (1, R g_R g_L^{-1} L^{-1})(L, L). \quad (1.11)$$

In the foregoing equations, the subscripts R, L refer to the decomposition $SU(3)_L \times SU(3)_R = SU(3)_V \times SU(3)_A$ as in the previous section. The last factor in eq. (1.11) can be discarded in G/H since $(L,L) \in SU(3)_V \sim H$. Therefore, every element of G/H can be represented by $U := g_R g_L^{-1}$, which transforms under the full group G according to

$$U(x) \to U'(x) = R U(x) L^\dagger, \quad L, R \in SU(3)_L, SU(3)_R. \quad (1.12)$$

Since we are dealing here with a local symmetry, U is a function of the space-time coordinates x. We still have the freedom to choose coordinates on the group manifold G/H. The matrix U is a unitary element of G/H and a function of the pseudoscalar Goldstone boson fields ϕ. It is convenient to choose the field variables such that

$$U = \exp\left(\frac{i\phi}{F}\right), \quad (1.13)$$

where the hermitean matrix ϕ is given by

$$\phi = \phi^a \lambda^a = \sqrt{2} \begin{pmatrix} \frac{1}{\sqrt{2}}\pi^0 + \frac{1}{\sqrt{6}}\eta & \pi^+ & K^+ \\ \pi^- & -\frac{1}{\sqrt{2}}\pi^0 + \frac{1}{\sqrt{6}}\eta & K^0 \\ K^- & \bar{K}^0 & -\frac{2}{\sqrt{6}}\eta \end{pmatrix}, \quad (1.14)$$

while F is a constant of mass dimension one, which in a further examination of the theory turns out to be the pion decay constant in the chiral limit, $F \sim 87$ MeV [20]. Of course, physical observables are always independent of the choice for the field parametrization (off-shell matrix elements will however depend on this choice in general).

We are now in a position to construct the effective Lagrangian describing the interaction among the Goldstone bosons. To guarantee simple transformation laws of the various building blocks, the Lagrangian is written in terms of the field matrix U which transforms linearly under G, see eq. (1.12). The effective Lagrangian is an infinite series of local terms involving the field U and multiple derivatives acting on it, as well as the external source fields v, a, s, p. Since we are attempting a low-energy expansion of the generating functional, Goldstone boson momenta are to be considered as being small compared to the hadronic scale ~ 1 GeV, usually referred to as the scale of chiral symmetry breaking and denoted by Λ_χ. The effective Lagrangian will therefore be ordered according to the number of derivatives contained in the respective terms: terms with higher numbers of derivatives are suppressed by factors of $O(q)$, where q generically stands for any small Goldstone boson momentum. The corresponding expansion of matrix elements in powers of q goes by the name of Chiral Perturbation Theory (ChPT) [8, 20, 21]. The suppression factors in higher-order terms turn out to be of the form $q/(4\pi F)$, where the denominator is numerically close to the hadronic scale Λ_χ. Disregarding the external sources for the moment, the only term in the effective Lagrangian with no derivatives is an unimportant constant proportional to some power of $U^\dagger U = 1$. There can be no term with only a single derivative because of Lorentz invariance (the Lagrangian must of course be a Lorentz scalar). At second order in the derivative expansion, there is a unique term

$$\mathcal{L}_{\text{eff}}^{(2)} = \frac{F^2}{4}\langle \partial_\mu U^\dagger \partial^\mu U \rangle + \ldots. \quad (1.15)$$

Here the brackets $\langle \cdots \rangle$ denote the trace in flavor space, and the dots stand for terms involving the external fields. The constant prefactor has been chosen such that the standard kinetic term for the scalar field ϕ is reproduced when expanding the exponential U, i.e.

$$\mathcal{L}_{\text{eff}}^{(2)} = \frac{1}{2}\partial_\mu \phi^a \partial^\mu \phi^a + O(\phi^4, v, a, s, p).$$

The field theory specified by eq. (1.15) is usually referred to as the non-linear sigma model. To make it invariant under local chiral transformations, we introduce a covariant derivative involving the external fields v and a:

$$\nabla_\mu U = \partial_\mu U - i(v_\mu + a_\mu)U + iU(v_\mu - a_\mu). \quad (1.16)$$

For consistency, the external fields v, a should thus be counted as $O(q)$, so that $\nabla_\mu U$ is of order q. Derivatives acting on the external fields will also produce factors of order q. For reasons that will become clear in a moment, the external scalar field s is counted as $O(q^2)$, so that it can only be included linearly in the second order effective Lagrangian. The appropriate invariant expression for $\mathcal{L}_{\text{eff}}^{(2)}$ including s is

$$\mathcal{L}_{\text{eff}}^{(2)} = \frac{F^2}{4}\langle \nabla_\mu U^\dagger \nabla^\mu U \rangle + \frac{F^2}{2}B\langle sU^\dagger + Us^\dagger \rangle, \quad (1.17)$$

where B is a constant and s has to transform as in eq. (1.7) for the last term to be invariant under chiral transformations. If we fix the external field at the value $s = \mathcal{M}$ (remember that $\mathcal{M}$

stands for the quark mass matrix), we effectively produce the explicit chiral symmetry breaking given by the quark mass terms of $\mathcal{L}_{\text{QCD}}$. The Lagrangian can thus be rewritten as

$$\mathcal{L}_{\text{eff}}^{(2)} = \frac{F^2}{4} \langle \nabla_\mu U^\dagger \nabla^\mu U \rangle + \frac{F^2}{2} B \langle \mathcal{M}(U + U^\dagger) \rangle. \tag{1.18}$$

Expanding the second term in powers of the Goldstone boson fields ϕ, one finds mass terms for the π,K and η mesons (first derived by Gell-Mann, Oakes and Renner [22]):

$$\begin{aligned}
M_{\pi^\pm}^2 &= 2B\hat{m}, \\
M_{\pi^0}^2 &= 2B\hat{m} + O\left(\frac{(m_u - m_d)^2}{m_s - \hat{m}}\right), \\
M_{K^\pm}^2 &= B(m_u + m_s), \\
M_{K^0}^2 &= B(m_d + m_s), \\
M_\eta^2 &= \frac{2}{3}B(\hat{m} + 2m_s) + O\left(\frac{(m_u - m_d)^2}{m_s - \hat{m}}\right),
\end{aligned} \tag{1.19}$$

where $\hat{m} = \frac{1}{2}(m_u + m_d)$. The corrections proportional to $(m_u - m_d)^2$ are related to π^0-η mixing. These terms are very small compared to the leading ones and can be neglected in most cases. On-shell Goldstone boson momenta satisfy $q^2 = M_\phi^2$, which in connection with the above mass formulae explains the counting rule $s \sim O(q^2)$. Of course, there will be corrections of higher orders in the quark masses to these mass relations. Letting the functional derivative $\delta/\delta s$ act on both sides of eq. (1.8) yields, to lowest order in momenta and quark masses,

$$\langle 0 \mid \bar{u}u \mid 0 \rangle = \langle 0 \mid \bar{d}d \mid 0 \rangle = \langle 0 \mid \bar{s}s \mid 0 \rangle = -F^2 B(1 + O(\mathcal{M})), \tag{1.20}$$

which expresses the a priori unknown constant B in terms of the scalar quark condensates $\langle 0 \mid \bar{q}q \mid 0 \rangle$ and the pion decay constant F. Note that we are employing here the scenario of standard ChPT, where B is treated as a large quantity of order one. For an alternative scenario, where the scalar quark condensates are assumed to be small parameters, we refer to [23]. As a further side-remark, we note that the mass relations of eqs. (1.19) yield the result

$$3M_\eta^2 = 4M_K^2 - M_\pi^2, \tag{1.21}$$

to leading order in the quark mass expansion, and in the isospin limit where $m_u = m_d$. Relation (1.21) is known as the Gell-Mann-Okubo mass formula and is experimentally found to be satisfied up to a few percent.

At lowest order, the effective Lagrangian method employing $\mathcal{L}_{\text{eff}}^{(2)}$ reproduces all results derived earlier using the method referred to as current algebra [24]. The method of effective Lagrangians is, however, much more general, because corrections to the lowest-order results can be derived in a systematic fashion in that framework. The Lagrangian is an infinite series of terms

$$\mathcal{L}_{\text{eff}} = \mathcal{L}_{\text{eff}}^{(2)} + \mathcal{L}_{\text{eff}}^{(4)} + \mathcal{L}_{\text{eff}}^{(6)} + \ldots$$

where the superscript refers to the 'chiral order' of the operators contained in the Lagrangian, i.e. the power of q associated to these operators, which is computed using the counting rules specified above. For later reference, we write down $\mathcal{L}_{\text{eff}}^{(4)}$, which was first constructed by Gasser and Leutwyler [20, 21]:

$$\mathcal{L}_{\text{eff}}^{(4)} = \sum_{i=1}^{10} L_i P_i + \sum_{i=11}^{12} H_i P_i, \quad (1.22)$$

where the operator structures P_i are given by

$$P_1 = \langle \nabla_\mu U^\dagger \nabla^\mu U \rangle^2, \qquad P_2 = \langle \nabla_\mu U^\dagger \nabla_\nu U \rangle \langle \nabla^\mu U^\dagger \nabla^\nu U \rangle,$$
$$P_3 = \langle \nabla_\mu U^\dagger \nabla^\mu U \nabla_\nu U^\dagger \nabla^\nu U \rangle, \qquad P_4 = \langle \nabla_\mu U^\dagger \nabla^\mu U \rangle \langle \chi^\dagger U + \chi U^\dagger \rangle,$$
$$P_5 = \langle \nabla_\mu U^\dagger \nabla^\mu U (\chi^\dagger U + U^\dagger \chi) \rangle, \qquad P_6 = \langle \chi^\dagger U + \chi U^\dagger \rangle^2,$$
$$P_7 = \langle \chi^\dagger U - \chi U^\dagger \rangle^2, \qquad P_8 = \langle \chi^\dagger U \chi^\dagger U + \chi U^\dagger \chi U^\dagger \rangle,$$
$$P_9 = -i \langle F_R^{\mu\nu} \nabla_\mu U \nabla_\nu U^\dagger + F_L^{\mu\nu} \nabla_\mu U^\dagger \nabla_\nu U \rangle, \qquad P_{10} = \langle U^\dagger F_R^{\mu\nu} U F_{\mu\nu}^L \rangle,$$
$$P_{11} = \langle F_{\mu\nu}^R F_R^{\mu\nu} + F_{\mu\nu}^L F_L^{\mu\nu} \rangle, \qquad P_{12} = \langle \chi^\dagger \chi \rangle.$$

In addition to previously defined building blocks, we have used here the abbreviations

$$\chi = 2B(s + ip), \qquad F_{R,L}^{\mu\nu} = \partial^\mu (v^\nu \pm a^\nu) - \partial^\nu (v^\mu \pm a^\mu) - i[v^\mu \pm a^\mu, v^\nu \pm a^\nu]. \quad (1.23)$$

The constants L_i and H_i are not constrained by the symmetries imposed on the effective Lagrangian. Rather, these so-called low-energy constants (LECs) are in principle determined by the effects of those degrees of freedom which are not explicitly contained in $\mathcal{L}_{\text{eff}}$. From the viewpoint of the effective field theory, the LECs are free parameters which have to be fixed using experimental input (there is, however, no guarantee that there exists enough experimental information so that this can always be done in practice for any LEC in question).

The number of LECs increases when going to higher chiral order. For example, $\mathcal{L}_{\text{eff}}^{(6)}$ is parametrized by 90 LECs in the three-flavor case [25]. There is no denying that, in general, the effective field theory will lose predictive power when considering higher-order effects. On the other hand, only a subset of operators will contribute to a certain process under consideration, so that it is not necessary to pin down all the LECs showing up in the effective Lagrangian at a given order.

As mentioned at the beginning of this section, we have restricted the symmetry group to $SU(3)_V \times SU(3)_A$, and we have dropped the singlet pieces of the external sources v, a. Otherwise, we would have to deal with anomalies (the axial $U(1)_A$ anomaly mentioned in the previous section, and the chiral anomaly [26, 27], related to the external singlet fields $\sim \text{tr } v_\mu$, $\text{tr } a_\mu$). In this more general case, we would have to include an additional piece in the fourth order Lagrangian, which accounts for the chiral anomaly. It was first given by Wess and Zumino [28] (see also [20, 29] for a discussion of these issues). Since we will only treat the non-anomalous sector of low-energy QCD in this work, we shall disregard these additional contributions in the following.

Now that we have discussed the effective Lagrangian, we must see how to set up the corresponding perturbation theory to compute all kinds of matrix elements. The power counting scheme for the effective Lagrangian can obviously be translated into a corresponding power counting for the vertices in Feynman graphs pertaining to this field theory. The propagators of the Goldstone boson fields ϕ^a read, in momentum space,

$$i\Delta(q) = \frac{i}{q^2 - M_\phi^2}$$

and should apparently be counted as $O(q^{-2})$. This rule will also be used when the corresponding internal line is part of a loop, where the loop momentum q is integrated up to infinity. We would like to demonstrate that this rule leads to a consistent counting scheme for Feynman diagrams using dimensional regularization. The simplest loop integral looks like

$$I_\phi = \int \frac{d^d q}{(2\pi)^d} \frac{i}{q^2 - M_\phi^2} \tag{1.24}$$

and is sometimes referred to as a 'tadpole-type' integral. Here d is the usual space-time dimension parameter of dimensional regularization. As an additional rule, the measure of the loop integration is counted as $O(q^d)$. The loop integral I_ϕ should therefore be of chiral order $d-2$. Indeed, one finds after the usual Wick rotation to Euclidean four-momenta that

$$I_\phi = M_\phi^{d-2} \frac{\Gamma(1-\frac{d}{2})}{(4\pi)^{\frac{d}{2}}} \tag{1.25}$$

(remember that M_ϕ is of order $O(q)$). It is crucial that we are working here with a mass-independent regularization scheme - otherwise, possible power divergences would spoil the counting rules for the loop graphs. To avoid this complication, we will be using dimensional regularization throughout in this work.

According to the above counting rules, a general Feynman graph with V_n vertices from $\mathcal{L}_{\text{eff}}^{(n)}$, I internal Goldstone boson lines and L loops will yield an amplitude $\mathcal{A}$ of chiral order $q^{D(\mathcal{A})}$, where the chiral dimension D is given by

$$D(\mathcal{A}) = dL - 2I + \sum_n n V_n. \tag{1.26}$$

Using the topological identity

$$L = I - \sum_n V_n + 1,$$

which simply expresses L as the number of momenta in the diagram that are not fixed by energy-momentum conservation, we find

$$D(\mathcal{A}) = (d-2)L + \sum_n (n-2) V_n + 2. \tag{1.27}$$

This already shows that, for $d \geq 2$, the contributions to the amplitude will be suppressed for an increasing number of loops. For definiteness, we will put $d = 4$ in the following. The lowest possible value of the chiral dimension is then given by $D(\mathcal{A}) = 2$, attributed to graphs with no loops (tree graphs) and an arbitrary number of vertex insertions from $\mathcal{L}_{\text{eff}}^{(2)}$. At next-to-leading order, there will be contributions from graphs with one loop and several vertices from the second-order Lagrangian, and tree graphs with only one insertion from $\mathcal{L}_{\text{eff}}^{(4)}$, and some insertions from $\mathcal{L}_{\text{eff}}^{(2)}$. At order q^6, two-loop graphs and vertices from $\mathcal{L}_{\text{eff}}^{(6)}$ enter, etc. It is clear from the above formula for $D(\mathcal{A})$ that, to every fixed chiral order, only a finite number of graphs will contribute.

Graphs with loops contain ultraviolet divergences for $d \to 4$ and require renormalization. A quantum field theory only makes sense if there are counterterms available which absorb these divergences[3]. This is not a problem for the effective field theory considered here: As long as we include all possible terms consistent with the required symmetries in the Lagrangian, every divergence generated by loop graphs derived from $\mathcal{L}_{\text{eff}}$ can be accounted for by a suitable counterterm derived from the same Lagrangian. Here we have assumed that a regularization scheme is used which maintains the symmetries of the theory. This is the case for dimensional regularization. For example, the divergences occuring in a one-loop calculation up to order $O(q^4)$ can always be compensated by renormalizing the LECs L_i in $\mathcal{L}_{\text{eff}}^{(4)}$, see eq. (1.22),

$$L_i = L_i^r(\mu) + k_i \bar{\lambda},$$

where the finite numerical coefficients k_i are given by the corresponding β-function, while the divergence is contained in

$$\bar{\lambda} = \frac{\mu^{d-4}}{16\pi^2}\left[\frac{1}{d-4} - \frac{1}{2}(\ln(4\pi) + 1 + \Gamma'(1))\right].$$

The subtraction of the term proportional to $\bar{\lambda}$ from the one-loop amplitudes specifies a certain scheme within dimensional regularization, called the modified minimal subtraction scheme ($\overline{MS}$ for short). We have also introduced an arbitrary mass parameter μ, known as the scale of dimensional regularization. The finite (renormalized) parts L_i^r depend on this scale, as well as the subtracted loop graphs. The full amplitude, including the loops and all possible counterterms up to a prescribed chiral order, is however independent of the choice for μ.

We have now collected all the tools necessary to start calculations in the meson sector of ChPT, where only the Goldstone bosons π,K,η are treated as active degrees of freedom. For many interesting applications in low-energy hadron physics, extensions of this framework are required. With the effective field theory constructed in this section, meson-baryon scattering at low energies, for example, can not be examined in a systematic way analogous to meson-meson scattering, where the above methods can be applied. A pure meson-field theory is simply not well suited for a treatment of processes where other particles like e.g. baryons play a major role. In the next two sections, we shall see how the method of evaluating chiral corrections in such more complicated situations can be systematized.

1.3 Baryon chiral perturbation theory

The aim of this section is to compute strong interaction matrix elements of the form

$$\mathcal{F}(\vec{p}', \vec{p}, v, a, s, p) = \langle B'(\vec{p}') \mid B(\vec{p}) \rangle^c_{v,a,s,p} \quad , \tag{1.28}$$

by means of a low-energy expansion, in analogy to the method explained in the previous section. B and B' denote the incoming and the outgoing baryon, with three-momenta $\vec{p}$ and $\vec{p}'$, respectively. The superscript c stands for 'connected', indicating that only connected Feynman diagrams contribute to $\mathcal{F}$. In close analogy to eq. (1.4), the above transition amplitude is evaluated in the presence of external sources v, a, s, p. We restrict ourselves here to the single-baryon sector, so that e.g. processes like baryon-baryon scattering are not in the scope of the

---

[3]The general theory of renormalization is most comprehensively reviewed in [30].

present section. More specifically, we include in our analysis only the baryon ground state octet of flavor $SU(3)$, parametrized in matrix form as

$$B = B^a \lambda^a = \begin{pmatrix} \frac{1}{\sqrt{2}}\Sigma^0 + \frac{1}{\sqrt{6}}\Lambda & \Sigma^+ & p \\ \Sigma^- & -\frac{1}{\sqrt{2}}\Sigma^0 + \frac{1}{\sqrt{6}}\Lambda & n \\ \Xi^- & \Xi^0 & -\frac{2}{\sqrt{6}}\Lambda \end{pmatrix}. \tag{1.29}$$

Under the vectorial subgroup $H = SU(3)_V$, B of course transforms as an octet. There are various possible choices for the transformation behaviour of B under the full chiral transformation group, corresponding to field redefinitions of the baryon fields. They all lead to the same physical results, though off-shell amplitudes in general depend on such choices of field variables. In the $SU(3)$-case, it is common use to require the following transformation law,

$$B \to B' = K(L, R, U)BK^\dagger(L, R, U), \tag{1.30}$$

where the so-called compensator field K is implicitly defined by the transformation law for the square root of U (written as u),

$$u \to u' = RuK^\dagger = KuL^\dagger, \qquad u^2 = U. \tag{1.31}$$

It may be seen that K is itself a unitary matrix. Remembering eqs. (1.5,1.6,1.12), it is straightforward to confirm that

$$u_\mu := iu^\dagger \nabla_\mu U u^\dagger \tag{1.32}$$

transforms according to $u_\mu \to Ku_\mu K^\dagger$ under local $SU(3)_L \times SU(3)_R$. Due to the covariant derivative entering in its definition, it is clear that u_μ (sometimes called the 'chiral vielbein') is of chiral order $O(q)$. It can be used as a building block in the lowest order effective meson-baryon Lagrangian. One finds [31, 32]

$$\mathcal{L}_{MB}^{(1)} = \langle \bar{B}(i\slashed{D} - m_0)B \rangle - \frac{D}{2}\langle \bar{B}\gamma^\mu \gamma_5 \{u_\mu, B\}\rangle - \frac{F}{2}\langle \bar{B}\gamma^\mu \gamma_5 [u_\mu, B]\rangle. \tag{1.33}$$

Here we introduced the covariant derivative

$$D_\mu B = \partial_\mu B + [\Gamma_\mu, B],$$
$$\Gamma_\mu = \frac{1}{2}(u^\dagger[\partial_\mu - i(v_\mu + a_\mu)]u + u[\partial_\mu - i(v_\mu - a_\mu)]u^\dagger), \tag{1.34}$$

which acts on fields transforming according to eq. (1.30). Moreover, m_0 denotes the common baryon octet mass in the chiral limit, and the coupling constants D and F can be determined from a fit to semileptonic baryon decays [33],

$$D \simeq 0.80, \qquad F \simeq 0.46.$$

Their sum is equal to the nucleon axial-vector coupling g_A in the chiral limit. Please note that the coupling F introduced here has nothing to do with the pion decay constant in the chiral limit.
The superscript (1) on the l.h.s. of eq. (1.33) refers to the fact that the displayed terms in the leading Lagrangian are of order $O(q)$. The kinetic term (the first term on the r.h.s. of

eq. (1.33)) is counted as $O(q)$, though the baryon mass m_0 is not small and counted as $O(1)$. This is because the baryon is pushed from its mass shell only by a small amount (of order $O(q)$) when interacting with low-energy Goldstone bosons. Further counting rules are

$$B, \bar{B} = O(1), \quad D_\mu B = O(1), \quad \bar{B}\gamma_5 B = O(q), \quad \bar{B}\gamma_\mu B = O(1), \quad \bar{B}\gamma_\mu\gamma_5 B = O(1).$$

Given the counting rules for the vertices from the meson-baryon Lagrangian, we can set up the ordering of the perturbation series in almost complete analogy to the one in the pure meson sector treated in the previous section. The only additional rule concerns the baryon propagators, which are counted as $O(q^{-1})$, as one would expect after the discussion of the kinetic term in the foregoing paragraph.

The effective field theory outlined above was worked out to one-loop order by Gasser, Sainio and Svarc [31], and will henceforth be referred to as baryon chiral perturbation theory (BChPT). Gasser et al. noted that, in contrast to the meson sector, dimensional regularization in combination with the $\overline{MS}$-scheme does not maintain the power counting for the loop graphs: In general, the loop integrals will yield terms of lower chiral order than one would expect from the naïve power counting. Therefore, strictly speaking, the ordering principle for the perturbation series breaks down in this scheme: There is no guarantee that a graph with dozens of loops will not contribute to the next-to-leading order result (or, for baryon masses, even to the leading term). This undesirable feature can be traced back to the appearance of an additional mass scale which is not of order $O(q)$, namely, the baryon mass in the chiral limit. Subsequently, various approaches were developed to circumvent this complication. The heavy-baryon formulation of BChPT [34, 35] provides a consistent power counting by means of a non-relativistic expansion in inverse powers of the baryon mass, at the expense of manifest Lorentz invariance, which is only restored perturbatively in the $(1/m_0)$-expansion. The nucleon momenta are split into a large piece of the order of the nucleon mass, and a small residual component which is counted as $O(q)$. This is achieved by a velocity-dependent projector, which also serves to divide the nucleon fields into a hard momentum component and a residual piece. The hard-momentum component is then integrated out in the path integral, the corresponding degrees of freedom being parametrized by an infinite series of local couplings in the new Lagrangian. The heavy-baryon method has been most widely used in BChPT - for extensive reviews, we refer to [36, 37]. One obvious drawback of this method is the loss of manifest Lorentz invariance. Moreover, the non-relativistic expansion of heavy-baryon ChPT in some cases fails to converge in parts of the low-energy region, see e.g. sec. 4.1 of [36], or sec. 3 of [38].

Taking up some earlier ideas of Ellis and Tang [39, 40], Becher and Leutwyler proposed a renormalization scheme [38] which has the advantage of maintaining manifest Lorentz invariance, and reproduces the correct analytic behaviour of the amplitudes in the low-energy region, while preserving the power counting rules. This scheme was called 'infrared regularization', and has in the meantime been applied to many interesting processes and observables. For a recent review, we again refer to [37]. We will show in this work that the original formulation of infrared regularization, designed for the treatment of the low-energy pion-nucleon system, can be extended to the case of the inclusion of explicit meson and baryon resonance fields in the effective Lagrangian. The corresponding extension of ChPT is reviewed in the next section. Variants of the infrared regularization scheme will be discussed in chapters 2 and 3.

1.4 Inclusion of resonance fields

It was already mentioned in sec. 1.2 that the effects induced by the degrees of freedom integrated out in the path integral of eq. (1.8) are encoded in a string of local couplings in the effective Lagrangians of (B)ChPT. It is also possible to include the resonance fields explicitly, as dynamical degrees of freedom in an effective Lagrangian, which is to be constructed following essentially the same lines as in the previous sections, namely, one must construct the most general Lagrangian including the resonance and Goldstone boson fields (and eventually the baryon fields) that is consistent with all required symmetry principles. Furthermore, one has to set up a power counting scheme to specify an ordering prescription for the perturbation series. Assume that we have done this. In the following, we restrict ourselves to the mesonic sector, but this restriction is not essential for the argumentation. Let $\mathcal{L}_{\text{Res}}(R, U, v, a, s, p)$ be our effective Lagrangian including the resonance fields R. Integrating out the resonances, we arrive at a generating functional Z_{ind} which contains a series of couplings induced by the resonances R,

$$\int [dR] \exp\left(i \int d^4 x \mathcal{L}_{\text{Res}}(R, U, v, a, s, p) \right) = e^{i Z_{\text{ind}}(U, v, a, s, p)}. \tag{1.35}$$

It is important to note that this will not induce any new structures in the effective Lagrangian describing the interaction among the Goldstone bosons, since all allowed operators are already included there. Thus the whole effect of the resonances is described by the contributions to the LECs of $\mathcal{L}_{\text{eff}}$ which are induced by the exchange of the resonances R, at least at low energies, where the effective theory can be expected to yield meaningful results. On the level of reaction amplitudes, this can be qualitatively illustrated as follows. The amplitude for two soft momentum Goldstone bosons interacting through an exchange of a resonance particle R will contain a pole at $t = M_R^2$, where t is the momentum transfer variable of the reaction, and M_R is the mass of the resonance. For small momenta, the pole structure can be expanded as a geometric series,

$$\frac{1}{M_R^2 - t} = \frac{1}{M_R^2} + \frac{t}{M_R^4} + \frac{t^2}{M_R^6} + \ldots, \quad t < M_R^2. \tag{1.36}$$

Therefore, at low energies, the resonance exchange is equivalent to a series of local couplings, producing terms of arbitary high order in the derivative expansion. Keeping the full pole structure in the amplitude therefore goes beyond the strict perturbative treatment of ChPT: it amounts to a resummation of higher order terms. The terms of a chiral order beyond the order one is working in the perturbative treatment can, strictly speaking, only be considered as an educated guess for the most important contributions at higher chiral order, since of course not all the higher order terms are determined in that way. Nonetheless, resummations have been shown to provide an improved description of experimental data in many cases. Resummations of the above kind, obtained by the inclusion of explicit resonance fields, can be particularly useful when the included resonances R are phenomenologically known to play an important role in the reaction under consideration. For example, it is well known for a long time that the rho meson plays an important role in the description of the electromagnetic form factor of the nucleon. The latter quantity has been calculated in [41], to order $O(q^4)$ in ChPT. In this work, it was also demonstrated that the inclusion of explicit vector meson degrees of freedom leads, through the resummation of higher order terms, to an improved description of the data in the low-energy region, providing a curvature effect of the theoretical prediction that could not be generated with the truncated perturbation series.

Chiral Lagrangians with resonances have also been used to examine the numerical values of the (renormalized) LECs appearing in the effective Lagrangians of (B)ChPT (as, for example, the coefficients L_i and H_i appearing in eq. (1.22)). With the above discussion in mind, it is quite natural to expect that the lowest-lying resonances not included in the respective Lagrangian are responsible for the main contributions to the pertinent LECs. In fact, this so-called principle of resonance saturation has been shown to be valid in the mesonic sector [42] as well as in the single-baryon sector [43]. Moreover, it turned out that, whenever the vector mesons can contribute, their contributions dominate those generated by the other (axial vector,(pseudo-)scalar...) meson resonances. This observation is summarized as vector meson dominance in phenomenological analyses. In the baryon sector, the $\Delta(1232)$ usually yields the most important contributions (for attempts to include the Δ-field explicitly in BChPT, we refer to [44–49]).

Motivated by the above-mentioned principle of vector meson dominance, we shall present, in the rest of this section, the treatment of the lowest-lying vector meson octet in the framework of ChPT. As an $SU(3)$ octet, the resonance fields collected in the field R transform as in eq. (1.30), $R \to K R K^\dagger$. Following [42,50] we employ an antisymmetric tensor field description of the spin-1 degrees of freedom, and write

$$R = W_{\mu\nu} = \frac{1}{\sqrt{2}} W_{\mu\nu}^a \lambda^a = \begin{pmatrix} \frac{\rho^0}{\sqrt{2}} + \frac{\omega}{\sqrt{2}} & \rho^+ & K^{*+} \\ \rho^- & -\frac{\rho^0}{\sqrt{2}} + \frac{\omega}{\sqrt{2}} & K^{*0} \\ K^{*-} & \bar{K}^{*0} & -\phi \end{pmatrix}_{\mu\nu}.$$

On the r.h.s. of this equation, we have written the resonance octet field in terms of the physical fields (assuming ideal $\phi - \omega$ mixing). The corresponding kinetic Lagrangian is

$$\mathcal{L}_W^{\text{kin}} = -\frac{1}{2} \langle D^\mu W_{\mu\nu} D_\rho W^{\rho\nu} \rangle + \frac{1}{4} M_V^2 \langle W_{\mu\nu} W^{\mu\nu} \rangle,$$

The brackets $\langle \ldots \rangle$ again denote the trace in flavor space, while the covariant derivative D_μ was defined in eq. (1.34). From $\mathcal{L}_W^{\text{kin}}$, one derives the tensor field propagator in momentum space,

$$T_{\mu\nu,\rho\sigma}^{ab}(q) = \frac{i\delta^{ab}}{M_V^2} \frac{g_{\mu\rho}g_{\nu\sigma}(M_V^2 - q^2) + g_{\mu\rho}q_\nu q_\sigma - g_{\mu\sigma}q_\nu q_\rho - (\mu \leftrightarrow \nu)}{M_V^2 - q^2}.$$

Since ChPT results are valid only for soft-momentum Goldstone bosons, we will restrict ourselves to processes where only those Goldstone bosons, or a single low-energy baryon, appear as external particles, while the vector mesons only appear as virtual low-energy particles, represented in Feynman diagrams as internal lines through which only a small momentum of order $O(q)$ is flowing. Remembering the expansion of the propagator denominator displayed in eq. (1.36), we conclude that the vector meson propagator $T(q)$ is to be counted as $O(1)$ in this situation, since the resonance mass M_V is $O(1)$. Moreover, derivatives acting on $W_{\mu\nu}$ will produce powers of soft momenta of order $O(q)$ in corresponding vertex rules.

At lowest chiral order, the interaction of the vector mesons with the Goldstone bosons is then given by [42]

$$\mathcal{L}_W^{\text{int}} = \frac{F_V}{2\sqrt{2}} \langle F_{\mu\nu}^+ W^{\mu\nu} \rangle + \frac{iG_V}{2\sqrt{2}} \langle [u_\mu, u_\nu] W^{\mu\nu} \rangle. \tag{1.37}$$

In the first term we have used the definition

$$F_{\mu\nu}^\pm = u F_{\mu\nu}^L u^\dagger \pm u^\dagger F_{\mu\nu}^R u, \tag{1.38}$$

as well as the definition of u_μ, see eq. (1.32). Recall that the external sources v_μ, a_μ are counted as $O(q)$, so that $F^\pm_{\mu\nu}$ is of chiral order $O(q^2)$. Also, u_μ is of $O(q)$. Therefore, $\mathcal{L}^{\text{int}}_W$ leads to vertices of chiral order $O(q^2)$.

Of course, at some stage of the calculation of matrix elements using the theory described above, one will also have to deal with loop graphs, with one or more vector meson internal lines being part of the loops. In this case, the counting for the resonance propagator does not seem to be justified, as the momentum flowing through the internal line is integrated up to infinity. In fact, loop integrals containing resonance propagators will not scale with the chiral order following from the naïve power counting rules. The reason is essentially the same as in BChPT, namely, the appearance of an additional mass scale (M_V in the present case). The loop integrals will pick up contributions from the region of the vector meson propagator pole, where the loop momentum ℓ is of order $\ell \sim M_V$, and those contributions will in general not be restricted by the low-energy counting rules. However, it turns out that the same terms can always be absorbed in a suitable renormalization of the LECs in the Lagrangian, utilizing an extended version of the infrared regularization scheme mentioned at the end of the previous section. This extended scheme was developed in [51], and will shortly be reviewed in the next chapter. Thus, the power counting for the graphs is well-defined up to a convenient renormalization procedure. In much the same way, the vector meson octet can be included explicitly in the single-baryon sector. The corresponding chiral Lagrangian at lowest order has been given in [52]. The extension of the power counting rules in this case is straightforward. The treatment of loop-graphs, utilizing a once more extended version of infrared regularization, will be explained in detail in the following chapter.

1.5 Unitarized chiral perturbation theory

From the viewpoint of the standard treatment of (B)ChPT, the inclusion of the effects of vector mesons to all orders in the expansion eq. (1.36) is a non-perturbative tool, the use of which should be justified by some underlying theoretical principle. In the mentioned case, it was the assumption that, beside the Goldstone bosons, the lowest-lying resonances yield the most important contributions to low-energy amplitudes, together with the phenomenologically motivated principle of vector meson dominance. In the present section, we shall explore the consequences of another guiding principle for the resummation of higher order terms, namely, exact two-body unitarity. Recall that unitarity of the $\mathcal{S}$-matrix, $\mathcal{S}^\dagger \mathcal{S} = 1$, is a fundamental requirement for any quantum field theory, since it expresses the 'conservation of probability' in any physical reaction, in the sense that the total probability that something will happen in the reaction is always equal to one. The number '1' in the unitarity statement for the $\mathcal{S}$-matrix stands for the identity operator in the Hilbert space of the quantum field theory. Writing the $\mathcal{S}$-matrix as $\mathcal{S} = 1 - i\mathcal{T}$, where $\mathcal{T}$ is called the reaction matrix, we can express the unitarity requirement, in operator form, as

$$\mathcal{T} - \mathcal{T}^\dagger = -i\mathcal{T}^\dagger \mathcal{T}. \tag{1.39}$$

It will be instructive to study the consequences of eq. (1.39) for a simple toy-model field theory, which will also be used in connection with gauge invariance in chapter 4. Consider the following Lagrangian for two complex scalar fields ϕ and ψ, with (bare) mass m and M, respectively:

$$\mathcal{L}_{\text{toy}} = \partial_\mu \phi^* \partial^\mu \phi - m^2 \phi^* \phi + \partial_\mu \psi^* \partial^\mu \psi - M^2 \psi^* \psi - g(\phi^* \phi)(\psi^* \psi). \tag{1.40}$$

At lowest order in a perturbation series in the coupling g, the scattering amplitude T for the process
$$\phi(q_1)\psi(p_1) \to \phi(q_2)\psi(p_2)$$
is simply given by the coupling g. At order g^2, one has to take one-loop graphs into account, etc. Truncating the perturbative expression for T at some finite order g^n, it is clear that the matrix elements
$$\langle \psi(p_2)\phi(q_2) | T | \phi(q_1)\psi(p_1) \rangle = (2\pi)^4 \delta^4(p_2 + q_2 - p_1 - q_1)T(s,t,u),$$
where s, t, u denote the usual Mandelstam variables
$$s = (p_1 + q_1)^2, \quad t = (p_2 - p_1)^2, \quad u = (p_2 - q_1)^2,$$
can not obey the unitarity relation eq. (1.39) exactly, but only up to corrections of $O(g^{n+1})$, since the quadratic expression $T^\dagger T$ on the r.h.s. of that relation contains terms of higher order than g^n, which are not present on the l.h.s. when inserting the truncated form of T. Thus, it is inevitable that any amplitude T that leads to an S-matrix which is exactly unitary will have to be constructed in a way that goes beyond perturbation theory.

In the following, we will further specialize to two-body unitarity, and therefore neglect intermediate states with more than two particles. Taking matrix elements of the unitarity relation for T, and inserting a complete set of two-particle states between the two T-operators on the r.h.s. of eq. (1.39), we find

$$\begin{aligned}&T(p_1 q_1 \to p_2 q_2) - T^*(p_2 q_2 \to p_1 q_1) \\ &= -i \int \frac{d^4q}{(2\pi)^4} \frac{d^4p}{(2\pi)^4} T^*(p_2 q_2 \to pq)\delta(q^2 - m^2) 2\pi\theta(q^0)\delta(p^2 - M^2) 2\pi\theta(p^0) T(p_1 q_1 \to pq) \\ &\quad \times (2\pi)^4 \delta^4(p_1 + q_1 - (p+q)). \end{aligned} \quad (1.41)$$

The integration here is nothing than the sum over the complete set of two-particle $\phi\psi$ states. The delta functions ensure energy-momentum conservation for the intermediate states, as well as the condition that the particles in the intermediate state are on their respective mass shell. The Heaviside step functions $\theta(\ldots)$ ensure that the intermediate particles are of positive energy. In going from eq. (1.39) to eq. (1.41), we have factored out an overall energy-momentum-conserving delta-function, so that $p_1 + q_1 = p_2 + q_2$ is understood in the latter equation.

Assume for the moment that the scattering amplitude T only depends on the Mandelstam variable s, so that it does not contain any dependence on the scattering angle (this would obviously be a pure s-wave scattering amplitude). The integration over the four-momenta p, q of the intermediate two-particle state can then be performed in eq. (1.41) to yield

$$T(s) - T^*(s) = -2iT^*(s)\Phi(s)T(s). \quad (1.42)$$

Here
$$\Phi(s) = \frac{|\boldsymbol{q}_{\text{cm}}|}{8\pi\sqrt{s}} \theta(s - (m+M)^2) \quad (1.43)$$
denotes the two-particle phase space factor, where
$$|\boldsymbol{q}_{\text{cm}}| = \frac{\sqrt{(s-(m+M)^2)(s-(m-M)^2)}}{2\sqrt{s}} \quad (1.44)$$

is the modulus of the three-momentum in the center-of-mass frame of the $\phi\psi$ particle system. The statement of two-body unitarity can be formulated in a very compact form,

$$\text{Im}(T^{-1}) = \Phi(s), \tag{1.45}$$

but remember that this simple statement is only valid under the assumption that the scattering amplitude depends only on the Mandelstam variable s. More generally, the simple condition eq. (1.45) will only apply for the s-wave part of the amplitude.
Every amplitude $T(s)$ which satisfies eq. (1.45) yields an $\mathcal{S}$-matrix which is exactly unitary in the subspace of two-particle states. A further requirement which we shall impose is that $T(s)$ agrees with the perturbative result on tree level, i.e. at lowest order in the coupling g. The simplest version for such an amplitude would be

$$\tilde{T}(s) = [g^{-1} + i\Phi(s)]^{-1}.$$

Such an amplitude has a very important shortcoming. The real and imaginary parts of scattering amplitudes are not independent of one another, but are linked by analyticity requirements [4], which can be expressed through so-called dispersion relations. A more refined version of a unitarized amplitude would be to include the full one-loop function

$$G(s) = \int \frac{d^d l}{(2\pi)^d} \frac{i}{((p_1 + q_1 - l)^2 - m^2 + i\epsilon)(l^2 - M^2 + i\epsilon)}, \tag{1.46}$$

which is however divergent for $d \to 4$. $G(s)$ can be written in a dispersive form, which makes its analyticity properties in the variable s manifest:

$$G(s) = G(s_0) - \frac{(s - s_0)}{\pi} \int_{s_{\text{thr}}}^{\infty} \frac{\Phi(s') ds'}{(s' - s)(s' - s_0)}, \quad d \to 4, \tag{1.47}$$

where $s_{\text{thr}} = (m + M)^2$ is the two-particle threshold value of the variable s, and s_0 is a complex number that does not lie on the integration contour $C = [s_{\text{thr}}, \infty]$, but can otherwise be chosen arbitrarily. The subtraction constant $G(s_0)$ contains the divergence as $d \to 4$. For physical values of s, $s \in C$, one has to approach the integration contour from above. This prescription is connected to the $i\epsilon$-terms displayed in the loop integral in eq. (1.46). Using this prescription, one finds Im $G(s) = -\Phi(s)$ for $s \in \mathbb{R}$. The interpretation of the representation eq. (1.47) is that G is an analytic function in the complex s-plane, except for branch points at s_{thr} and infinity, so that the corresponding branch cut can be chosen along the contour C. The discontinuity across the cut is given by $2i\Phi(s)$. Including the full loop function G in T, we restore some of the analyticity properties which the scattering amplitude should have from general field-theoretical considerations. So let us write our ansatz as

$$T(s) = [g^{-1} - G(s)]^{-1} = [\bar{g}^{-1}(s_0) - D(s, s_0)]^{-1}. \tag{1.48}$$

Here $D(s, s_0) = G(s) - G(s_0)$ is given by the (finite) dispersion integral on the r.h.s. of eq. (1.47), and we have introduced a 'renormalized' coupling constant

$$\bar{g}(s_0) = [g^{-1} - G(s_0)]^{-1}, \tag{1.49}$$

---
[4]The analytic properties of the $\mathcal{S}$-matrix are reviewed in the textbook [53]. Though one usually invokes the principle of causality to derive those analyticity constraints, it is possible to base essentially the same arguments mainly on Lorentz invariance, see e.g. the discussion in sections 5.1 and 10.8 of [3].

which is equal to $T(s_0)$. This value must be fixed by some further input, which can be seen as an analogue of a renormalization condition. Having fixed the value of $\bar{g}(s_0)$, all the other values of $T(s)$ are direct predictions of the model amplitude eq. (1.48).

It is important to note that the unitarized solution $T(s)$ still has shortcomings and is not consistent with all fundamental constraints from field theory. One such constraint is given by crossing symmetry, which in the present case demands that the scattering amplitude be symmetric under the interchange of Mandelstam variables $s \leftrightarrow u$. Moreover, though we have included the full analytic function G in our ansatz, the behaviour of $T(s)$ could in principle still be at odds with the strictures of analyticity, because these amplitudes can in general have poles on the first (physical) Riemann sheet in the variable s, which should be absent according to fundamental field-theoretic considerations. The constraint of the absence of such poles can be used to restrict the values of the couplings (like $\bar{g}$) in the unitarized amplitude, see e.g. the discussion in chapter 2 of [54].

Despite these shortcomings, unitarized amplitudes, similar in form to eq. (1.48), have been widely used in ChPT, in the meson as well as in the baryon-sector [55–58]. Unitarization procedures can be particularly useful when resonances are located in the vicinity of low-energy thresholds. In such cases, the scattering amplitude can not be reproduced using the standard perturbative treatment of ChPT, since the radius of convergence of the perturbation series becomes too small due to the nearby presence of the resonance pole (a prominent example is given by the $\Lambda(1405)$ resonance in K^-p-scattering [59]). It is clear from the general structure of unitarized amplitudes as e.g. in eq. (1.48) that resonance-like structures can be produced in this non-perturbative framework, owing to the possible presence of zeroes in the denominator of such amplitudes, which should be located on the second Riemann sheet for genuine resonances. In that case the resonance is said to be 'dynamically generated' in the unitarized approach, so that one has a clue to examining the nature of the resonance under consideration. In fact, the model amplitudes of unitarized ChPT have been used to predict resonance poles in the low-energy regime. However, one should always keep in mind that these amplitudes are not derived with the same theoretical rigour as standard ChPT results. Namely, chiral symmetry, in combination with a strict power counting, is not respected by unitarized amplitudes, and can only be restored perturbatively by matching those amplitudes to the perturbative result, up to some prescribed chiral order.

There are various methods of unitarization, but the corresponding scattering amplitudes will all be more or less of the form $[K^{-1} + i\Phi(s)]^{-1}$, differing only in the choice of K^{-1}, which may be a real function or a hermitian matrix, depending on the number of channels one includes in the scattering amplitude (in the latter case, the phase space function $\Phi(s)$ is of course also extended to a matrix in the space of reaction channels). In the analysis of the toy model presented above, we have already anticipated a justification for our choice of Re T^{-1}, which basically amounts to the inclusion of the full loop function $G(s)$ for the sake of analyticity. This choice has a further advantage, which will turn out to be important in chapters 4 and 5, when implementing the constraint of gauge invariance: It allows us to find a direct connection of our unitarized amplitude to the framework of Feynman diagrams usually employed in perturbation theory. To see this on a formal level, we return once more to our toy model, eq. (1.40). In this model, the unitarized amplitude constructed in eq. (1.48) is easily seen to be a solution of the following integral equation, referred to as the Bethe-Salpeter equation (BSE) for the scattering amplitude:

$$\begin{aligned}T(p_1q_1 \to p_2q_2) &= V(p_1q_1 \to p_2q_2) \\ &+ \int \frac{d^d l}{(2\pi)^d} T(l,(P-l) \to p_2q_2) \frac{i}{((P-l)^2 - m^2)(l^2 - M^2)} V(p_1q_1 \to l,(P-l)),\end{aligned}$$
(1.50)

where $P = p_1 + q_1 = p_2 + q_2$, and where the potential V is given in the present case simply by g. Indeed, under the assumption that the amplitude T depends only on $P^2 = s$, the BSE reduces to an algebraic equation, which immediately yields the result of eq. (1.48). The interpretation in terms of Feynman graphs is then given as an infinite sum of 'bubble' graphs, in which the potential V is iterated (see also fig. 1.1). This interpretation is of course already obvious when $T(s)$ is formally expanded in a Taylor series in the coupling g,

$$T(s) = g + gG(s)g + gG(s)gG(s)g + \ldots,$$

but the point here is that the solution of the BSE is also defined when the above expansion in g is not valid, e.g. when s approaches a resonance pole.

In Unitarized Chiral Perturbation Theory (UChPT), the potential V that is to be iterated in the BSE must be derived from the relevant chiral Lagrangian. To avoid double counting of contributions, the potential must correspond to the sum of two-particle irreducible graphs, truncated at a given order in perturbation theory. In general, the potential is momentum-dependent, and it is then a non-trivial task to solve the integral equation (1.50). Sometimes, the BSE is directly reduced to an algebraic equation by restricting V to the on-shell potential. This procedure is referred to as the 'on-shell approximation'. It is usually justified argueing that the off-shell parts of the potential lead to terms proportional to tadpole-type loop integrals (see e.g. eq. (1.24)) which can be absorbed in the couplings contained in V. Since this approximation clearly deviates from the correct field-theoretic treatment of Feynman graphs, we prefer to solve the BSE with the full potential as derived from the corresponding Lagrangian. However, this leads to yet another difficulty. The off-shell behaviour of the vertices depends on the choice of the field variables in the Lagrangian, and is in general ambiguous. Therefore, such off-shell dependence must drop out in the final result, when it comes to calculating predictions for physical observables. In the present non-perturbative approach, only a certain subset of Feynman diagrams, namely, the sum of bubble chain graphs, is selected from the full set of diagrams contributing to the scattering process. Consequently, there is no guarantee that the off-shell behaviour of vertices will drop out in the resulting scattering amplitude. So, strictly speaking, the chiral Lagrangian in connection with the BSE leads to a whole class of unitarized amplitudes T, parametrized by the possible choices for the field variables. The argument we have just cited as a justification for the use of the on-shell approximation implies that the ambiguity related to the off-shell dependence of vertices is reflected in the freedom to adjust the coupling parameters appearing in the potential. This freedom can in turn be related to the appearance of the arbitrary scale μ showing up in renormalized loop integrals, or equivalently to the choice of the subtraction point s_0 as e.g. in eqs. (1.48) and (1.49). This brings us to the issue of renormalization. The momentum integration to be performed in the BSE contains UV-divergences as $d \to 4$. In our toy model amplitude eq. (1.48), we have absorbed the resulting divergence of the solution T in the bare coupling g. It will be shown in app. F that a similar procedure is possible for the chiral meson-baryon potential that we use in chapters 4 and 5. Though it is

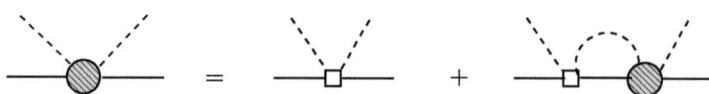

Figure 1.1: Graphical illustration of the BSE for the toy-model field theory. The filled circle represents the scattering amplitude T, while the open square stands for the vertex given by the coupling g.

reassuring that one can deal with the occuring divergences by absorbing them in the potential, it is important to note that this is not directly related to a rigorous field-theoretical treatment employing counter terms in the Lagrangian. In fact, every counter term vertex derived from a Lagrangian will show crossing symmetry, while the amplitudes generated by the BSE necessarily violate this symmetry, since the potential is only iterated along the s-channel. To our knowledge, a rigorous way for a non-perturbative renormalization of the BSE does not exist yet. For a more detailed discussion of the problems concerning off-shell dependence and UV-divergences in the BSE-approach to UChPT, we refer to [60]. Details on the solution of the BSE for the lowest order meson-baryon potential will be given in chapter 5 and app. E.

Chapter 2

Infrared regularization with vector mesons and baryons

2.1 Introduction

In a recent work [51], the scheme of infrared regularization (developed in its original form by Becher and Leutwyler [38]) was extended to the case of explicit meson resonances interacting with soft pions, as e.g. the first step towards a systematic inclusion of vector mesons in the meson sector of Chiral Perturbation Theory (ChPT). In this chapter, we will provide a further extension of infrared regularization to the situation where baryons as well as vector mesons are present. This can be considered as a synthesis of the results in [38] and [51]. Such an extension is not only of interest in itself, but can be applied to a plethora of observables, where vector and axial-vector mesons are known to play an important role, such as the electroweak form factors of the nucleon. As one example we mention the contribution of the $\pi\rho$ loop to the strangeness form factors of the nucleon [61,62].

Infrared regularization (IR) is a solution to the following problem. We have already mentioned in sec. 1.3 that the presence of mass scales which must be considered as 'heavy' compared to the masses and momenta of the soft pions will in general mess up the usual power counting rules of ChPT by which the perturbation series of the effective theory is ordered [8,20,21]. This observation was first made when baryons were incorporated in the framework of ChPT [31]. The procedure of infrared regularization separates the (dimensionally regularized) one-loop graphs of baryon ChPT into a part which stems from the soft pion contribution and a part generated from loop momenta close to the 'heavy' scale. The latter portion of the loop graph, called the 'regular' part, will usually not be in accord with the low-energy power counting, but can always be absorbed in local terms derived from the effective Lagrangian. It is therefore dropped from the loop graph, and only the first part, called the infrared singular part of the loop integral, is kept. Though both the vector mesons and the baryons interact as 'heavy' particles with the pions, the power counting is different for the two species, at least for the kinds of Feynman graphs we consider in this work. There, the vector mesons appear only as internal lines with small momenta far from their mass shell, so that the resonance propagator is counted as $O(q^0)$ (where q indicates some small momentum scale or Goldstone boson mass), while the baryon propagator is counted as $O(q^{-1})$, since the baryon is pushed from its mass shell only by a small amount due to its interaction with the soft pions and vector mesons. In the graphs we treat here, only one single baryon is present, with the baryon line running through the diagram undergoing only soft interactions. The number of the virtual vector mesons, however, is not fixed. The pion propagator is counted as $O(q^{-2})$, as usual, both the pion momentum and the pion mass being of $O(q)$. Appropriate powers of q are also assigned to vertices from the effective Lagrangian. Finally, the measure of every d-dimensional loop integration is booked as $O(q^d)$. This counting scheme applies, of course, to tree graphs, but also to the infrared singular, or soft, parts of the loop graphs. The regular parts of the loop graphs are not guaranteed to obey the power counting rules. These general remarks will be exemplified in the following sections. Before working out the case where both baryons and vector mesons appear in a Feynman diagram, we will briefly review the scheme of infrared regularization for loop integrals where only one heavy scale shows up. This will not only serve to give a unified presentation of the method, but also provide some results needed for an application of the general scheme to the axial form factor of the nucleon. In this and the following chapter, we focus on the $SU(2)$ case, in view of the applications treated in these chapters. The method outlined here can be trivially extended to the $SU(3)$ framework. Before starting with the presentation of the formalism, let us mention that the loop integrals studied in this work have also been treated, using a different regularization scheme which is in some respect complementary to the one used here, in ref. [63].

2.2 IR regularization in the pion-nucleon system

When only pions and nucleons are treated as explicit fields of the effective theory, the fundamental loop integral one has to consider is

$$I_{MB}(p^2) = \int \frac{d^d l}{(2\pi)^d} \frac{i}{((p-l)^2 - m^2)(l^2 - M^2)}. \tag{2.1}$$

Here, M is the pion mass (being of chiral order $O(q)$) and m is the nucleon mass. Applying the low-energy power counting scheme outlined in the introduction, one would assign a chiral order of q^{d-3} to this integral. We will see in a moment that only an appropriately extracted low-energy part of I_{MB} will obey this power counting requirement.

All the other pion-nucleon loop integrals are either only trivially modified by the infrared regularization scheme, or they can be derived from eq. (2.1) (see sec. 6 of [38]). For example, the scalar tadpole integral containing only the pion propagator is not modified at all, as there is no 'hard momentum' structure present that could lead to a nonvanishing regular part of this integral. Thus we have

$$I_M^{IR} = I_M = \int \frac{d^d l}{(2\pi)^d} \frac{i}{l^2 - M^2},$$

and a direct calculation gives

$$I_M^{IR} = \frac{\Gamma(1 - \frac{d}{2})}{(4\pi)^{\frac{d}{2}}} M^{d-2}. \tag{2.2}$$

Note the typical structure of the d-dependent power of the pion mass. For arbitrary values of the dimension parameter d, I_M^{IR} is in general proportional to fractional powers of M^2, and will even diverge in the so-called chiral limit where the quark masses m_u, m_d go to zero (so that also $M \to 0$, see e.g. eq. (1.19)) for small enough d. Such terms will never occur in the regular parts of the loop integrals which stem from the high-momentum region of the integration: those parts are always expandable in the pion mass. The baryon tadpole integral, e.g., is

$$I_B = \int \frac{d^d l}{(2\pi)^d} \frac{i}{l^2 - m^2},$$

and is trivially expandable in M^2, since it has no pion mass dependence at all. Therefore one has $I_B^{IR} = 0$. These remarks may serve to explain the terminology 'infrared singular' vs. 'regular'.

Returning to eq. (2.1), we must extract the part of this integral that is proportional to some d-dependent power of M, like in eq. (2.2). One can think of this extraction prescription as an operational definition of infrared regularization. The method is explained in full generality in [38]. Here, we concentrate on the case with on-shell momentum p, i.e. $p^2 = m^2$. This shows all the features we need for the demonstration, and also yields the result we will use in our application of the scheme in sec. 2.4. We introduce a Feynman parameter integration in the usual way:

$$I_{MB} = \int \frac{d^d l}{(2\pi)^d} \int_0^1 \frac{idz}{[((p-l)^2 - m^2)z + (l^2 - M^2)(1-z)]^2}. \tag{2.3}$$

Performing the standard steps, we find for $p^2 = m^2$:

$$I_{MB} = -m^{d-4} \frac{\Gamma(2 - \frac{d}{2})}{(4\pi)^{\frac{d}{2}}} \int_0^1 \frac{dz}{(z^2 - \alpha z + \alpha)^{2-\frac{d}{2}}}, \tag{2.4}$$

where we have defined $\alpha = M^2/m^2$ (note the difference to ref. [38], where this letter is reserved for $\alpha_{BL} = M/m$). Fractional powers of the small variable α will be produced near $z = 0$: there, the integrand is approximately $(\alpha)^{\frac{d}{2}-2}$. For small enough d, there would be an infrared singularity for $M \to 0$ located in parameter space at $z = 0$. It can already be seen from eq. (2.3) that the parameter region near $z = 0$ is associated with the low-energy portion of the integral: In this region, only the 'soft' pion propagator is weighted in the loop integration, while the hard momentum structure of the nucleon propagator dominates near $z = 1$. The extraction of the part of the integral proportional to d-dependent powers of α now proceeds as follows: the parameter integration is split into two parts like

$$\int_0^1 = \int_0^\infty - \int_1^\infty. \tag{2.5}$$

We will first show that the second integral on the r.h.s. is regular, i.e. expandable, in the variable α. This is easy to see, because for $z \geq 1$, the integrand can be expanded like

$$(z^2 - \alpha z + \alpha)^{\frac{d}{2}-2} = z^{d-4} \sum_{k=0}^\infty \frac{\Gamma(2 - \frac{d}{2} + k)}{\Gamma(2 - \frac{d}{2})} \frac{\alpha^k}{k!} \left(\frac{z-1}{z^2}\right)^k. \tag{2.6}$$

Interchanging integration and summation (which is a valid operation at least for some range of d), one gets for the regular part

$$R \equiv m^{d-4} \frac{\Gamma(2 - \frac{d}{2})}{(4\pi)^{\frac{d}{2}}} \int_1^\infty \frac{dz}{(z^2 - \alpha z + \alpha)^{2-\frac{d}{2}}} = m^{d-4} \frac{\Gamma(2 - \frac{d}{2})}{(3 - d)(4\pi)^{\frac{d}{2}}} + O(\alpha). \tag{2.7}$$

At this point we should make the remark that the extension of the parameter integration to infinity will lead to divergences as d increases. The infrared singular or regular parts are then defined as follows: The parameter integrals are computed for the range of d where they are well-defined, and the result will be continued analytically to arbitrary values of d. This amounts to the suppression of power divergences of the parameter integrals, which have nothing to do with the infrared singularity at $z = 0$, and will be cancelled anyway on the r.h.s. of eq. (2.5). Next we must show that the first term on the r.h.s. of eq. (2.5) is proportional to a d-dependent power of α. To see this, we substitute $z = \sqrt{\alpha} y$ to get

$$\int_0^\infty \frac{dz}{(z^2 - \alpha z + \alpha)^{2-\frac{d}{2}}} = \sqrt{\alpha}^{d-3} \int_0^\infty \frac{dy}{(y^2 - \sqrt{\alpha} y + 1)^{2-\frac{d}{2}}}.$$

The remaining integral on the r.h.s. will not produce d-dependent powers of α, since the integrand can be expanded in $\sqrt{\alpha}$ similar to eq. (2.6). Thus we have found that the parameter integral from zero to infinity is proportional to a d-dependent power of α. The infrared singular part I_{MB}^{IR} of the loop integral therefore equals

$$\begin{aligned}
I_{MB}^{IR} &= -m^{d-4} \frac{\Gamma(2-\frac{d}{2})}{(4\pi)^{\frac{d}{2}}} \int_0^\infty \frac{dz}{(z^2 - \alpha z + \alpha)^{2-\frac{d}{2}}} \\
&= -\frac{m^{d-4} \sqrt{\alpha}^{d-3}}{2(4\pi)^{\frac{d}{2}}} \sum_{k=0}^\infty \frac{\sqrt{\alpha}^{-k}}{k!} \Gamma\left(\frac{k+1}{2}\right) \Gamma\left(\frac{3+k-d}{2}\right).
\end{aligned} \quad (2.8)$$

As $d \to 4$, this has the well-known leading term

$$I_{MB}^{IR}(d \to 4) = \frac{1}{16\pi}\left(\frac{M}{m}\right) + \dots,$$

where the dots indicate terms of higher order in M/m. This is clearly in accord with the low-energy power counting. In contrast to that, the first term of the expansion of the regular part R obviously violates this counting when $d \to 4$. The infrared regularization now prescribes to drop R from the loop contribution and substitute I_{MB}^{IR} for I_{MB}. Moreover, the poles of I_{MB}^{IR} in $d-4$ are also absorbed in a renormalization of the masses and coupling constants of the effective Lagrangian. Again, for a more general and comprehensive treatment of the IR scheme in the pion-nucleon system, the reader should consult the original article of Becher and Leutwyler [38].

2.3 IR regularization for vector mesons and pions

In this section, we consider another case of infrared regularization, first examined in [51]. The internal baryon lines from the preceding section are now replaced by vector meson lines, however, the vector mesons do not show up as external particles in the graphs we consider here (an example for the treatment of such a graph can be found in sec. 11 of [51]). The only external particles here are pions (or, in some cases, soft photons etc.), so that there are only small external momenta of order $O(q)$ flowing into the loop. The fundamental scalar loop integral is in this case

$$I_{MV}(q^2) = \int \frac{d^d l}{(2\pi)^d} \frac{i}{((q-l)^2 - M_V^2)(l^2 - M^2)}, \quad (2.9)$$

where M_V is the mass of the heavy meson resonance and q is some small external momentum (small with respect to the resonance mass M_V). This is, in principle, the same function as in eq. (2.1), so the reader might ask why we devote this section to the examination of this case. The point is that the extraction of the infrared singular part of I_{MV} must proceed along different lines here. As shown in [51], a splitting like that of eq. (2.5) does not amount to a separation into infrared singular and regular parts as in the preceding section. This can be traced back to the fact that the extension of the parameter integrals to infinity leads to a singular behaviour of the loop function near $q^2 = 0$. Of course, also I_{MB}^{IR} has such a singularity as the external momentum squared goes to zero (see sec. 5.4 of [38]), but the point $p^2 = 0$ lies far outside the low-energy region in that case. Here, however, the point $q^2 = 0$ lies at the center of the low-energy region, so the infrared singular and regular parts can not be expressed as parameter integrals from zero or one to infinity, respectively. For example, a 'regular' part defined as being proportional to the parameter integral from one to infinity would not be expandable around $q^2 = 0$, as it should be in order to be able to absorb the corresponding terms in a renormalization of the local operators in the effective Lagrangian.

To circumvent this difficulty, we will take up the following simple idea from [51]. We observe that I_{MV} is analytic at $q^2 = 0$, the only singularity being the threshold branch point at $q^2 = (M_V + M)^2$, which is far outside the range of small q^2-values. Expanding I_{MV} in q^2, the analyticity properties in that variable are obvious. Each coefficient in this expansion can be split into an infrared singular and a regular part almost like in eq. (2.5). Extracting the infrared singular part proportional to d-dependent powers of the pion mass of each coefficient, and resumming the series, one arrives at a well-defined expression for the infrared singular part of the loop integral I_{MV} that is (by construction) expandable in the small variable q^2/M_V^2.

To start with the analysis, let us first consider the special case where the external momentum vanishes: $q = 0$. Then we have

$$I_{MV}(0) = \int \frac{d^d l}{(2\pi)^d} \frac{i}{(l^2 - M_V^2)(l^2 - M^2)}$$

$$= \frac{1}{M^2 - M_V^2} \left(\int \frac{d^d l}{(2\pi)^d} \frac{i}{l^2 - M^2} - \int \frac{d^d l}{(2\pi)^d} \frac{i}{l^2 - M_V^2} \right).$$

In the last step of this equation, we have already achieved the splitting into an infrared singular and a regular part, which is quite trivial here. The first term is proportional to a d-dependent power of M (compare eq. (2.2)), while the second part is clearly expandable in M^2 (because $M_V^2 \gg M^2$). Moreover, the pion mass expansion of the first term starts with M^{d-2}, in agreement with the low-energy power counting for I_{MV}: In contrast to the baryon propagator in sec. 2.2, the resonance propagator is counted as $O(q^0)$ since, in the low-energy region, the off-shell momentum of the internal resonance line is far below its mass shell. The counting for the pion propagator and the loop measure is the same as before, and one is lead to the power counting result q^{d-2} for I_{MV}. Therefore we can write

$$I_{MV}^{IR}(0) = \frac{1}{M^2 - M_V^2} \int \frac{d^d l}{(2\pi)^d} \frac{i}{l^2 - M^2}. \tag{2.10}$$

It should be clear that there will only be corrections of $O(q^2)$ to this result when we compute the regularized integral for nonvanishing q. We introduce the following variables,

$$\tilde{\alpha} = \frac{M^2}{M_V^2}, \quad \tilde{\beta} = \frac{q^2}{M_V^2}, \tag{2.11}$$

which we assume to be small in the sense that $\tilde{\alpha}, \tilde{\beta} \ll 1$. Just like in sec. 2.2, we use the Feynman parameter trick to write

$$I_{MV}(q^2) = -M_V^{d-4} \frac{\Gamma(2-\frac{d}{2})}{(4\pi)^{\frac{d}{2}}} \int_0^1 \frac{dz}{(\tilde{\beta}z^2 + z(1-\tilde{\alpha}-\tilde{\beta}) + \tilde{\alpha})^{2-\frac{d}{2}}}. \tag{2.12}$$

We rewrite this expression as follows:

$$I_{MV}(q^2) = -M_V^{d-4} \frac{\Gamma(2-\frac{d}{2})}{(4\pi)^{\frac{d}{2}}} \int_0^1 \frac{dz}{(z(1-\tilde{\alpha}-\tilde{\beta})+\tilde{\alpha})^{2-\frac{d}{2}}} \left(1 + \frac{\tilde{\beta}z^2}{z(1-\tilde{\alpha}-\tilde{\beta})+\tilde{\alpha}}\right)^{\frac{d}{2}-2}.$$

This can be expanded according to

$$I_{MV}(q^2) = -\frac{M_V^{d-4}}{(4\pi)^{\frac{d}{2}}} \sum_{k=0}^{\infty} \frac{\Gamma(\frac{d}{2}-1)}{k!\Gamma(\frac{d}{2}-1-k)} \int_0^1 \frac{\Gamma(2-\frac{d}{2})dz}{(z(1-\tilde{\alpha}-\tilde{\beta})+\tilde{\alpha})^{2-\frac{d}{2}}} \left(\frac{\tilde{\beta}z^2}{z(1-\tilde{\alpha}-\tilde{\beta})+\tilde{\alpha}}\right)^k.$$

In the next step, we extract the infrared singular part of each term in the sum. In each parameter integral, we substitute $z = \tilde{\alpha}y$ and extend the integration range to infinity:

$$I_k \equiv \int_0^1 dz \frac{z^{2k}}{(z(1-\tilde{\alpha}-\tilde{\beta})+\tilde{\alpha})^{2+k-\frac{d}{2}}} \to \int_0^\infty dy \frac{\tilde{\alpha}^{\frac{d}{2}-1+k}y^{2k}}{(1+y(1-\tilde{\alpha}-\tilde{\beta}))^{2+k-\frac{d}{2}}}. \tag{2.13}$$

Divergences of the parameter integral due to the extension of the upper limit to infinity are again handled by analytic continuation in d. It can be expressed in terms of Gamma functions:

$$\int_0^\infty dy \frac{y^{2k}}{(1+y(1-\tilde{\alpha}-\tilde{\beta}))^{2+k-\frac{d}{2}}} = \frac{\Gamma(2k+1)\Gamma(1-\frac{d}{2}-k)}{(1-\tilde{\alpha}-\tilde{\beta})^{2k+1}\Gamma(2-\frac{d}{2}+k)}.$$

This is clearly expandable in the small variables $\tilde{\alpha}$ and $\tilde{\beta}$, so that the r.h.s. of eq. (2.13) in fact has the proper form of an infrared singular contribution, being proportional to d-dependent powers of $\tilde{\alpha}$. Moreover, it is not difficult to see that the parameter integrals from 1 to infinity are completely regular in the small variables. So, putting pieces together, and using the following identity for Gamma functions:

$$\frac{\Gamma(\frac{d}{2}-1)}{\Gamma(\frac{d}{2}-1-k)} = (-1)^k \frac{\Gamma(2-\frac{d}{2}+k)}{\Gamma(2-\frac{d}{2})}, \quad k \in \mathbb{N}, \tag{2.14}$$

we can sum the series of infrared singular terms and write

$$I_{MV}^{IR}(q^2) = -\frac{M_V^{d-4}}{(4\pi)^{\frac{d}{2}}} (\tilde{\alpha})^{\frac{d}{2}-1} \sum_{k=0}^{\infty} \frac{(-\tilde{\alpha}\tilde{\beta})^k}{(1-\tilde{\alpha}-\tilde{\beta})^{2k+1}} \frac{\Gamma(2k+1)\Gamma(1-\frac{d}{2}-k)}{\Gamma(k+1)}. \tag{2.15}$$

Note that the leading term in this result is of chiral order $O(q^{d-2})$, as predicted by the power counting scheme. All the terms we have separated off from I_{MV} are regular in both small parameters, and only the infrared singular terms remain. For $\tilde{\beta} = 0$, the result for $I_{MV}^{IR}(0)$, which was already established in eq. (2.10), is reproduced.

It is perhaps worth noting that the result of eq. (2.15) can be obtained in a different way, which goes back to Ellis and Tang [39, 40]. Though they use their method in the pion-nucleon sector, a variant of it is also applicable here. Their prescription to obtain the soft momentum contribution of a loop integral is the following: Expand the propagators of the heavy particles as if the loop momentum were small, and then interchange summation and loop integration. It is claimed that this prescription eliminates the 'hard momentum' contributions present in the full loop graph. If this is true, and the concept of hard vs. soft momentum effects is a well-defined one, the result of the procedure should reproduce the infrared singular part of the loop integral in question. Let us see how this works out for our example. Following the prescription just described step by step, we make the following set of transformations:

$$I_{MV}(q^2) \to \int \frac{d^d l}{(2\pi)^d} \frac{i}{l^2 - M^2} \sum_{k=0}^{\infty} \frac{(2q \cdot l)^k}{(q^2 + l^2 - M_V^2)^{k+1}}$$

$$\to \int \frac{d^d l}{(2\pi)^d} \frac{i}{l^2 - M^2} \sum_{k=0}^{\infty} \frac{(2q \cdot l)^k}{(q^2 + M^2 - M_V^2)^{k+1}}$$

$$\to \sum_{k=0}^{\infty} \int \frac{d^d l}{(2\pi)^d} \frac{i(2q \cdot l)^k}{(l^2 - M^2)(q^2 + M^2 - M_V^2)^{k+1}}.$$

While the first step is just the expansion of the vector meson propagator pole imposed by the prescription, the second step deserves a comment: There, we have used the same trick of partial fractions to split off some hard momentum contributions as in the treatment of $I_{MV}(0)$, but now this was performed $k+1$ times. In the last step, summation and integration were interchanged. Using the formula

$$\int \frac{d^d l}{(2\pi)^d} \frac{i(q \cdot l)^{2n}}{l^2 - M^2} = (-q^2 M^2)^n M^{d-2} \frac{\Gamma(n + \frac{1}{2})}{\Gamma(\frac{1}{2})} \frac{\Gamma(1 - \frac{d}{2} - n)}{(4\pi)^{\frac{d}{2}}} \quad (2.16)$$

and the fact that the loop integrals in the series vanish if k is odd, the result of the transformation is

$$I_{MV}^{\text{soft}}(q^2) = \frac{M^{d-2}}{(4\pi)^{\frac{d}{2}}} \sum_{n=0}^{\infty} \frac{(-4q^2 M^2)^n}{(q^2 + M^2 - M_V^2)^{2n+1}} \frac{\Gamma(n + \frac{1}{2})\Gamma(1 - \frac{d}{2} - n)}{\Gamma(\frac{1}{2})}. \quad (2.17)$$

Extracting a factor of M_V^{d-4}, and using the identity

$$\frac{\Gamma(2n+1)}{\Gamma(n+1)} = 4^n \frac{\Gamma(n + \frac{1}{2})}{\Gamma(\frac{1}{2})}, \quad (2.18)$$

we see by comparing to eq. (2.15) that $I_{MV}^{\text{soft}} = I_{MV}^{IR}$.

In [51], the result for I_{MV}^{IR} was given in a different form (see eqs. (7.10) and (8.4) of that reference). Of course, it is equivalent to the result derived above: it is shown in app. A that the two different forms just amount to a reordering of the corresponding expansions. We will find that the form of eq. (2.15) is most practical for our purposes, in particular, the chiral expansion can almost immediately be read off from that formula.

In closing this section, we note that the spin or the parity of the resonance obviously do not play a major role in the above considerations. Though we will concentrate on the case of vector mesons, most of what we have said would also apply for other meson resonances, like e.g. scalar or axial-vector mesons.

2.4 Pion-nucleon system with explicit meson resonances

Now that we have collected the results for the fundamental one-loop integrals in the pion-nucleon and the vector meson-pion sector, we are prepared to extend the framework of infrared regularization once more, and apply it to Feynman graphs where nucleons, pions as well as vector mesons take part in the same loop. The simplest example where this situation occurs is the triangle graph consisting of one pion, one nucleon and one vector meson line. Due to baryon number conservation, the nucleon line must run through the complete diagram. We shall assume that only a small momentum k (small in the usual sense) is transferred at the vector meson-pion vertex. Such a graph will typically contribute to some nucleon form factor in the region of small momentum transfer. With this application in mind, and with the excuse that it will simplify the presentation a bit, we will further specify to on-shell nucleons. Let p and $\bar{p}$ be the four-momenta of the incoming and the outgoing nucleon. Then we have

$$p^2 = m^2 = \bar{p}^2 = (p+k)^2 \Rightarrow k^2 = 2\bar{p} \cdot k = -2p \cdot k. \qquad (2.19)$$

Using these kinematic relations, we can rewrite the fundamental scalar loop integral

$$I_{MBV}(k^2) \equiv \int \frac{d^d l}{(2\pi)^d} \frac{i}{((p-l)^2 - m^2)((k+l)^2 - M_V^2)(l^2 - M^2)} \qquad (2.20)$$

with the help of the usual Feynman parameter trick, as

$$\int \frac{d^d l}{(2\pi)^d} \int_0^1 \int_0^{1-y} \frac{2i\,dxdy}{(y((p-l)^2 - m^2) + x((k+l)^2 - M_V^2) + (1-x-y)(l^2 - M^2))^3}. \qquad (2.21)$$

Doing the loop integration in the usual manner, we get

$$I_{MBV}(k^2) = m^{d-6} \frac{\Gamma(3-\tfrac{d}{2})}{(4\pi)^{\tfrac{d}{2}}} \int_0^1 \int_0^{1-y} \frac{dxdy}{(y^2 - \alpha y + \alpha + \beta(x^2 + xy) + x(\gamma - \alpha - \beta))^{3-\tfrac{d}{2}}}, \qquad (2.22)$$

where the definitions

$$\alpha = \frac{M^2}{m^2} \ll 1, \qquad \beta = \frac{k^2}{m^2} \ll 1, \qquad \gamma = \frac{M_V^2}{m^2} \sim O(1).$$

were used. In particular, we have assumed here that the nucleon and the meson resonance are roughly of the same order of magnitude (in the real world, we have $\gamma \sim 2/3$ for the rho resonance, which is good enough for our purposes).

We should remark here that there is, of course, a second graph with the same topology, where the pion and the resonance line are interchanged (see fig. 2.2). The expression for the corresponding scalar loop integral is

$$\tilde{I}_{MBV}(k^2) \equiv \int \frac{d^d l}{(2\pi)^d} \frac{i}{((\bar{p}-l)^2 - m^2)((l-k)^2 - M_V^2)(l^2 - M^2)}. \qquad (2.23)$$

However, the reader can convince himself that, due to the on-shell kinematics specified in eq. (2.19), this will give exactly the same expression as in eq. (2.22). Thus, we can focus on the integral I_{MBV}.

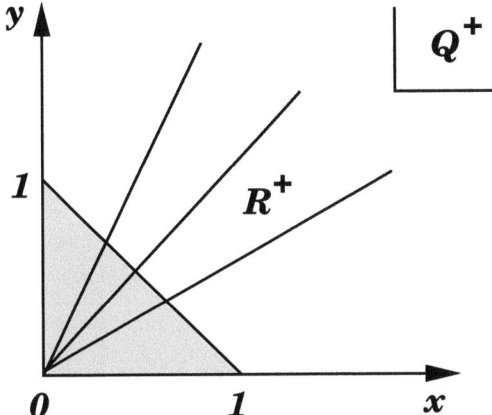

Figure 2.1: Illustration of the integration ranges for the infrared singular and the regular part of I_{MBV}. The shaded triangle is the region integrated over in eq. (2.22). The three rays indicate the extension of the connections from the soft point at $(0,0)$ to the hard line from $(0,1)$ to $(1,0)$.

In analogy to sec. 2.3, it will be instructive to begin with the special case where $k = 0$. Since we must split off the terms where only propagators of heavy particles occur, we can obviously apply the same partial fraction method that led to eq. (2.10). The remaining pion-nucleon integral can be dealt with as in sec. 2.2. This gives

$$I^{IR}_{MBV}(0) = \frac{I^{IR}_{MB}(m^2)}{M^2 - M_V^2}. \tag{2.24}$$

Having the standard method described in sec. 2.2 in mind, we note that this equals

$$I^{IR}_{MBV}(0) = m^{d-6} \frac{\Gamma(3 - \frac{d}{2})}{(4\pi)^{\frac{d}{2}}} \int_0^\infty \int_0^\infty \frac{dxdy}{(y^2 - \alpha y + \alpha + x(\gamma - \alpha))^{3-\frac{d}{2}}}$$

$$= \frac{m^{d-6}}{(\gamma - \alpha)} \frac{\Gamma(2 - \frac{d}{2})}{(4\pi)^{\frac{d}{2}}} \int_0^\infty \frac{dy}{(y^2 - \alpha y + \alpha)^{2-\frac{d}{2}}}$$

(compare the last line with the l.h.s of eq. (2.8)). Once again, the possible divergence for large d at $x \to \infty$ was regularized by analytic continuation from small d as before. It is reassuring to see that eq. (2.24) is reproduced in this way.

The extension of both parameter integrations to infinity is the natural generalization of the prescription used in sec. 2.2. This can be seen as follows. In eq. (2.21), we chose our Feynman parameters such that the pion propagator contributes with the weight one at $x = y = 0$, while the vector meson and the nucleon propagator have their maximum weight at $(x, y) = (1, 0)$ and $(0, 1)$, respectively. Consequently, the infrared singularity is located in parameter space at the point $(x, y) = (0, 0)$. Indeed, a look at eq. (2.22) confirms that the integrand in that expression is approximately $\alpha^{\frac{d}{2}-3}$ in the region where $x \sim y \sim 0$, thus producing infrared singular terms there. Far from this region, the integrand is expandable in α, as we will show later. To be more specific, imagine the positive quarter of the parameter plane, i.e. $Q_+ \equiv \{(x,y) : x \geq 0, y \geq 0\}$, split in two parts, the first being the triangle one integrates over in the full loop integral (see eq. (2.22)) and the second part its complement in Q_+, named R_+ (see fig. 2.1 for an illustration). In the latter parameter region, no infrared singularities are located, hence, the integration over this region should yield a proper regular part (there is a qualification here, see below). Geometrically, one might imagine that all lines from the 'soft point' $(0,0)$ (the location of the infrared singularity in parameter space) to the 'hard line' from $(0,1)$ to $(1,0)$ are extended to infinity to get the infrared singular part of the loop integral. In a more general setting, there may be soft or hard points, lines, surfaces etc., depending on the number of soft and hard pole structures in the integral under consideration. Extending all connecting lines from the soft point, line etc. to the hard point (line...) to infinity, one always achieves a splitting in the original parameter region and a region where no infrared singularities in parameter space are present, analogous to R_+. It is possible to show that this geometric picture leads to the same results as the prescription given in sec. 6 of [38]. This picture might serve as a guide when trying to split arbitrary one-loop graphs into infrared singular and regular parts, however, one should always convince oneself that both those parts have the correct properties, and that their sum equals the original integral. One has to be a bit careful here, as the following qualification shows. As explained in sec. 2.3, extending the integration range to infinity might lead to unphysical singularities in variables in which the original integral is analytic (at least in the low-energy region). Those singularities will be harmless if they are located far from the low-energy region, as e.g. the singularity at $s=0$ of the infrared singular parts in the pion-nucleon sector, but in other cases, they can be disturbing. Following the method of [51], we avoided this problem in sec. 2.3 by expanding the original loop integral in the small variable $\tilde{\beta}$ beforehand. Only then could the integration range be extended to infinity, in each coefficient of the expansion. We must expect that a similar phenomenon will occur in the present case.

To exclude this from the start, we expand I_{MBV} in analogy to eq. (2.15):

$$I_{MBV}(k^2) = \frac{m^{d-6}}{(4\pi)^{\frac{d}{2}}} \sum_{j=0}^{\infty} \frac{\Gamma(\frac{d}{2}-2)}{\Gamma(\frac{d}{2}-2-j)} \int_0^1 \int_0^{1-y} \frac{\Gamma(3-\frac{d}{2}) dx dy (\beta x(x+y))^j}{j!(y^2 - \alpha y + \alpha + x(\gamma - \alpha - \beta))^{3-\frac{d}{2}+j}}$$

$$= \frac{m^{d-6}}{(4\pi)^{\frac{d}{2}}} \sum_{j=0}^{\infty} \frac{\Gamma(\frac{d}{2}-2)}{\Gamma(\frac{d}{2}-2-j)} \int_0^1 \int_0^{1-y} \frac{\Gamma(3-\frac{d}{2}) dx dy (\sum_{l=0}^{j} \binom{j}{l} \beta^j x^{l+j} y^{j-l})}{j!(y^2 - \alpha y + \alpha + x(\gamma - \alpha - \beta))^{3-\frac{d}{2}+j}}.$$

There is still some β-dependence in the denominator, but in that combination, it will turn out to be harmless. We extend the parameter integrations to Q_+ and define

$$I_{MBV}^{IR}(k^2) = \frac{m^{d-6}}{(4\pi)^{\frac{d}{2}}} \sum_{j=0}^{\infty} \frac{\Gamma(\frac{d}{2}-2)}{\Gamma(\frac{d}{2}-2-j)} \int_0^{\infty} \int_0^{\infty} \frac{\Gamma(3-\frac{d}{2}) dx dy (\sum_{l=0}^{j} \binom{j}{l} \beta^j x^{l+j} y^{j-l})}{j!(y^2 - \alpha y + \alpha + x(\gamma - \alpha - \beta))^{3-\frac{d}{2}+j}}.$$

Since the denominator is always positive for $0 < \alpha \ll 1, |\beta| \ll 1$, the x-integration can readily be done using the formula

$$\int_0^{\infty} dx \frac{x^n}{(a+bx)^D} = \frac{a^{n+1-D}}{b^{n+1}} \frac{\Gamma(n+1)\Gamma(D-(n+1))}{\Gamma(D)}, \tag{2.25}$$

together with eq. (2.14). This gives

$$I^{IR}_{MBV}(k^2) = \frac{m^{d-6}}{(4\pi)^{\frac{d}{2}}} \sum_{j=0}^{\infty} \sum_{l=0}^{j} \frac{\Gamma(j+l+1)\Gamma(2-\frac{d}{2}-l)}{\Gamma(j-l+1)\Gamma(l+1)} \frac{(-\beta)^j}{(\gamma-\alpha-\beta)^{j+l+1}} \times$$
$$\times \int_0^{\infty} \frac{y^{j-l}dy}{(y^2-\alpha y+\alpha)^{2-\frac{d}{2}-l}}.$$

In the next step, we can write down the chiral expansion of the y-integral, using the same method as in sec. 2.2. The generalized formula is

$$\int_0^{\infty} dy \frac{y^n}{(y^2-\alpha y+\alpha)^D} = \sqrt{\alpha}^{n+1-2D} \sum_{k=0}^{\infty} \frac{\sqrt{\alpha}^k}{k!} \frac{\Gamma(\frac{n+k+1}{2})\Gamma(\frac{2D+k-(n+1)}{2})}{2\Gamma(D)}. \quad (2.26)$$

Obviously, the chiral expansion of this integral can straightforwardly be read off from the series on the r.h.s. Inserting this result, we get

$$I^{IR}_{MBV}(k^2) = \frac{m^{d-6}}{(4\pi)^{\frac{d}{2}}} \sum_{j,k=0}^{\infty} \sum_{l=0}^{j} \frac{(-\beta)^j \sqrt{\alpha}^{d-3+j+l+k}}{(\gamma-\alpha-\beta)^{j+l+1}} \frac{\Gamma(j+l+1)\Gamma(\frac{j+k-l+1}{2})\Gamma(\frac{3-d-l-j+k}{2})}{2\Gamma(j-l+1)\Gamma(k+1)\Gamma(l+1)}. \quad (2.27)$$

The only expressions we have not expanded in the small variables are the factors of $(\gamma-\alpha-\beta)$ in the denominator, but this can of course be done: the corresponding geometric series is absolutely convergent due to the assumption that $\gamma \sim O(1)$.
To complete the proof that eq. (2.27) is the correct infrared singular part of I_{MBV}, we have to show that all the terms we dropped in the extraction procedure described above are regular in α. Those terms are proportional to parameter integrals over the region R_+, of the general type

$$R_j = \int_{R_+} \frac{dxdy(x(x+y))^j}{(y^2-\alpha y+\alpha+x(\gamma-\alpha-\beta))^{3-\frac{d}{2}+j}}$$
$$= \int_{z=1}^{\infty} \int_{x=0}^{z} \frac{dxdz(xz)^j}{((z-x)^2-\alpha(z-1)+x(\gamma-\beta))^{3-\frac{d}{2}+j}}.$$

Here we have traded the variable y for $z \equiv x+y$. One finds that the function

$$f(z,x) = \frac{(z-x)^2+x(\gamma-\beta)}{z-1}$$

has the property

$$f(z,x) \geq \min\{4,(\gamma-\beta)\}$$

in R_+, provided that the parameters β and γ are in their typical low-energy ranges. This is already sufficient to ensure that the integrand of R_j can safely be expanded in $(z-1)\alpha$, and that integration and summation of the corresponding series can be interchanged. This proves the regularity of the integrals R_j in α.
It is also possible to show that the result for I^{IR}_{MBV} can be obtained in a way that is closely analogous to the prescription of Ellis and Tang (see the end of sec. 2.3). The proof can be found in app. B.

2.5 Application: Axial form factor of the nucleon

A typical example for an application where the triangle graph treated in sec. 2.4 shows up is a contribution to some form factor of the nucleon at low momentum transfer. As a specific example, we consider the nucleon form factor of the isovector axial-vector current in a theory with an explicit rho resonance field. A representation for this form factor, using the infrared regularization scheme in the pion-nucleon sector, has been given by Schweizer [64]. In that framework, the contributions due to the various baryon or meson resonances are contained in the low-energy coefficients (LECs) of the effective pion-nucleon Lagrangian, or, more correctly: The contributions from tree-level resonance exchange can be described as an infinite sum of contact terms derived from the pion-nucleon Lagrangian. The inclusion of explicit resonance fields therefore amounts to a resummation of higher order terms, which is often advantageous (as discussed in detail in ref. [41]). Moreover, a theory with explicit resonance fields can serve to achieve an understanding of the numerical values of the LECs, relying on the assumption that the lowest-lying resonances give the dominant contributions to those coefficients. This is usually called the principle of resonance saturation, and has been very successfull in ChPT, see e.g. [42, 43]. Assuming this principle to be valid, the pion-nucleon LECs can be expressed through the masses of the resonances and the couplings of the resonances to the nucleons and pions. Such relations are most useful, of course, if the resonance masses and couplings are sufficiently well known. In this section, we will compute the contribution to the axial form

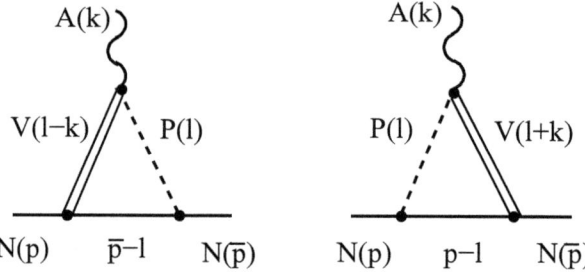

Figure 2.2: Triangle graphs contributing to the axial form factor of the nucleon. The dashed line represents the pion, while the double line stands for the vector meson. The letter A indicates the external axial source, N denotes the nucleon under consideration.

factor of the nucleon that is described by the two triangle graphs of fig. (2.2), with one nucleon, one pion and one resonance line, in the framework of the extended infrared regularization scheme developed in the preceding sections. First, we must set up the necessary formalism and collect the various terms from the effective Lagrangian we need for the computation.
By Lorentz invariance, the matrix element of the axial current

$$A^i_\mu(x) \equiv \bar{q}(x)\gamma_\mu\gamma_5\frac{\tau^i}{2}q(x)$$

between one-nucleon states can be parametrized as

$$\langle N'(\bar{p})|A^i_\mu(x)|N(p)\rangle = \bar{u}'(\bar{p})\left(G_A(t)\gamma_\mu + G_P(t)\frac{k_\mu}{2m} + G_T(t)\frac{\sigma_{\mu\nu}k^\nu}{2m}\right)\gamma_5\frac{\tau^i}{2}u(p)e^{ikx}. \qquad (2.28)$$

In the above expressions, q is the quark field spinor, $q^T = (u,d)$, N' is the outgoing nucleon with momentum $\bar{p}$, N labels the incoming nucleon with momentum p, and $k = \bar{p} - p$, $t \equiv k^2$. The symbols τ^i denote the usual Pauli matrices. Finally, $\bar{u}'$ and u are the Dirac spinors associated with the outgoing and incoming nucleon, respectively. Assuming perfect isospin symmetry and charge conjugation invariance, as we will do here, leads to $G_T \equiv 0$. The relation of G_A and G_P to the quantities $F_{1,2}$ used in [64] is $F_1(t) = G_A(t), 2mF_2(t) = -G_P(t)$. For later reference, we give the representation of G_A up to order q^3 that can be found in [64]:

$$G_A(t) = g_A + 4\bar{d}_{16}M^2 + d_{22}t - \frac{g_A^3 M^2}{16\pi^2 F^2}$$
$$+ \frac{g_A M^3}{24\pi m F^2}\left(3 + 3g_A^2 - 4c_3 m + 8c_4 m\right) + O(q^4). \qquad (2.29)$$

Here F and g_A denote the pion decay constant and the nucleon axial charge in the chiral limit, respectively, while the coefficients c_i, d_i are LECs showing up in the pion-nucleon effective Lagrangian at order two and three, respectively. For a precise definition of the underlying Lagrangian see ref. [65].

We now turn to the calculation of the axial form factor in the presence of vector mesons. We write down the relevant terms in the effective Lagrangian and give the necessary rules for the vertices and propagators required for the calculation. First, the lowest order chiral Lagrangian for the pion-nucleon interaction is derived from eq. (1.33) by projection on the $SU(2)$ sector (i.e. by neglecting all fields except for the pions and nucleons),

$$\mathcal{L}_N^{(1)} = \bar{\psi}(i\slashed{D} - m)\psi + \frac{g_A}{2}\bar{\psi}\slashed{u}\gamma_5\psi. \qquad (2.30)$$

Here, ψ is the nucleon spinor, $\psi^T = (p,n)$, and $g_A = D + F$. The matrix u_μ collects the pion fields π^a via

$$u = \exp\left(\frac{i\vec{\tau}\cdot\vec{\pi}}{2F}\right),$$
$$u_\mu = i\{u^\dagger, \partial_\mu u\} + u^\dagger r_\mu u - u l_\mu u^\dagger,$$
$$r_\mu = v_\mu + a_\mu, \quad l_\mu = v_\mu - a_\mu.$$

In the last line, we have introduced external isovector vector and axial-vector sources, v_μ and a_μ,

$$v_\mu = v_\mu^i \frac{\tau^i}{2}, \quad a_\mu = a_\mu^i \frac{\tau^i}{2}.$$

The covariant derivative D_μ was defined in eq. (1.34). In the $SU(2)$ framework, it can be written as

$$D_\mu = \partial_\mu + \Gamma_\mu,$$
$$\Gamma_\mu = \frac{1}{2}[u^\dagger, \partial_\mu u] - \frac{i}{2}u^\dagger r_\mu u - \frac{i}{2}u l_\mu u^\dagger.$$

From eq. (2.30), one derives the $\bar{N}N\pi$ vertex rule
$$\frac{g_A}{2F}\not{q}\gamma_5\tau^a$$
for an outgoing pion of momentum q.

Now we turn to the effective Lagrangians involving the vector meson fields. We choose a representation in terms of an antisymmetric tensor field $W_{\mu\nu}$ [42, 50, 51]. The corresponding free Lagrangian is
$$\mathcal{L}_W^{\text{kin}} = -\frac{1}{2}\langle D^\mu W_{\mu\nu} D_\rho W^{\rho\nu}\rangle + \frac{1}{4}M_V^2\langle W_{\mu\nu} W^{\mu\nu}\rangle,$$
where
$$W_{\mu\nu} = \frac{1}{\sqrt{2}} W_{\mu\nu}^i \tau^i = \begin{pmatrix} \frac{\rho^0}{\sqrt{2}} & \rho^+ \\ \rho^- & -\frac{\rho^0}{\sqrt{2}} \end{pmatrix}_{\mu\nu}.$$

The brackets $\langle\ldots\rangle$ denote the trace in isospin space. From $\mathcal{L}_W^{\text{kin}}$, one derives the tensor field propagator in momentum space,
$$T_{\mu\nu,\rho\sigma}^{ij}(k) = \frac{i\delta^{ij}}{M_V^2} \frac{g_{\mu\rho}g_{\nu\sigma}(M_V^2-k^2) + g_{\mu\rho}k_\nu k_\sigma - g_{\mu\sigma}k_\nu k_\rho - (\mu\leftrightarrow\nu)}{M_V^2 - k^2}.$$

At lowest chiral order, the interaction of the rho meson with the pions is given by the Lagrangian cited in eq. (1.37),
$$\mathcal{L}_W^{\text{int}} = \frac{F_V}{2\sqrt{2}}\langle F_{\mu\nu}^+ W^{\mu\nu}\rangle + \frac{iG_V}{2\sqrt{2}}\langle[u_\mu, u_\nu]W^{\mu\nu}\rangle. \tag{2.31}$$

We also remind the reader of the following definitions introduced in chapter 1,
$$F_{\mu\nu}^\pm = u F_{\mu\nu}^L u^\dagger \pm u^\dagger F_{\mu\nu}^R u,$$
$$F_{\mu\nu}^L = \partial_\mu l_\nu - \partial_\nu l_\mu - i[l_\mu, l_\nu],$$
$$F_{\mu\nu}^R = \partial_\mu r_\nu - \partial_\nu r_\mu - i[r_\mu, r_\nu].$$

As explained in sec. 1.4, the $\mathcal{L}_W^{\text{int}}$ leads to vertices of chiral order $O(q^2)$. Using the method of external sources to derive Greens functions from the generating functional, we must extract the amplitudes linear in the source a_μ to compute the matrix element of eq. (2.28). For the triangle graphs considered here, we need the vertex that connects the vector meson and the pion with the external axial source. From eq. (2.31), we find the corresponding vertex rule
$$\epsilon^{iac}\left(\frac{G_V}{F}(q_\mu g_{\nu\tau} - q_\nu g_{\mu\tau}) - \frac{F_V}{2F}(k_\mu g_{\nu\tau} - k_\nu g_{\mu\tau})\right).$$

Here i, a, c are the isospin indices associated with the axial source a_τ, the pion and the vector meson field $W_{\mu\nu}$, respectively, q is the four-momentum of the outgoing pion. Since k and q are counted as small momenta, the chiral order of this vertex rule is in accord with the power counting for the interaction Lagrangian.

The restrictions of chiral symmetry are not that strong for the interaction of the vector mesons with the nucleons: Here, there are terms of chiral order $O(q^0)$. The leading terms of the interaction Lagrangian have been given in ref. [52]. For the $SU(2)$ case we consider here, the relevant terms are

$$\mathcal{L}_{NW} = R_V \bar{\psi} \sigma_{\mu\nu} W^{\mu\nu} \psi + S_V \bar{\psi} \gamma_\mu D_\nu W^{\mu\nu} \psi$$
$$+ T_V \bar{\psi} \gamma_\mu D_\lambda W^{\mu\nu} D^\lambda D_\nu \psi + U_V \bar{\psi} \sigma_{\lambda\nu} W^{\mu\nu} D^\lambda D_\mu \psi. \quad (2.32)$$

In the notation of [52], we have $R_V = R_D + R_F$, $S_V = S_D + S_F$, etc. The definition of $\sigma_{\mu\nu}$ is standard, $\sigma_{\mu\nu} = i[\gamma_\mu, \gamma_\nu]/2$. It turns out that only the piece proportional to G_V from eq. (2.31) and the piece proportional to R_V from eq. (2.32) contribute at lowest order to the diagrams computed here, the chiral expansion of which starts at $O(q^3)$ (given that a scheme like infrared regularization is used that preserves the power counting rules). To keep the presentation short, we show only the contribution from those terms and therefore neglect some higher order contributions. However, this will be sufficient to compare our results to the representation up to $O(q^3)$ given by Schweizer [64].

The evaluation of the first graph gives (see fig. (2.2))

$$I_1 = \int \frac{d^d l}{(2\pi)^d} \left(\frac{G_V}{F} (l_\mu g_{\nu\tau} - l_\nu g_{\mu\tau}) \epsilon^{iac} \right) \frac{i}{l^2 - M^2} \left(-\frac{g_A}{2F} \slashed{l} \gamma_5 \tau^a \right) \frac{i T_{cd}^{\mu\nu,\rho\sigma}(l-k)}{\slashed{p} - \slashed{l} - m} i\sigma_{\rho\sigma} \tau^d \frac{R_V}{\sqrt{2}},$$

and the second one gives

$$I_2 = \int \frac{d^d l}{(2\pi)^d} \frac{R_V}{\sqrt{2}} i\sigma_{\mu\nu} \tau^c \frac{i T_{cd}^{\mu\nu,\rho\sigma}(l+k)}{\slashed{p} - \slashed{l} - m} \left(\frac{g_A}{2F} \slashed{l} \gamma_5 \tau^a \right) \frac{i}{l^2 - M^2} \left(\frac{G_V}{F} (l_\rho g_{\sigma\tau} - l_\sigma g_{\rho\tau}) \epsilon^{ida} \right).$$

We have left out the Dirac spinors $\bar{u}, u$ here. Since we consider on-shell nucleons, we can use the Dirac equation to simplify the numerators of the integrals, $\bar{u}(\bar{p})\slashed{\bar{p}} = \bar{u}(\bar{p})m$, $\slashed{p} u(p) = mu(p)$. We will make some remarks on the computation of I_1 (the computation of I_2 can be done analogously). In a first step, we reduce the full loop integral to a linear combination of scalar loop integrals, which have been treated in detail in the preceding sections. Scalar loop integrals without a pion propagator denominator are dropped using infrared regularization, since they are pure regular parts. Therefore we can replace $l^2 \to M^2$ everywhere in the numerator. With the abbreviation

$$g_1 = \frac{2\sqrt{2} g_A G_V R_V}{M_V^2 F^2}$$

we get

$$I_1 = g_1 \tau^i \bigg(\int \frac{d^d l}{(2\pi)^d} \frac{2im\slashed{l}((M^2 - k \cdot l)(l^\rho - k^\rho)\sigma_{\rho\tau} + \frac{i}{2}(k_\tau - l_\tau)(\slashed{l}\slashed{k} - \slashed{k}\slashed{l}))\gamma_5}{((\bar{p}-l)^2 - m^2)((l-k)^2 - M_V^2)(l^2 - M^2)}$$
$$+ \int \frac{d^d l}{(2\pi)^d} \frac{i((M^2 - k \cdot l)(l^\rho - k^\rho)\sigma_{\rho\tau} + \frac{i}{2}(k_\tau - l_\tau)(\slashed{l}\slashed{k} - \slashed{k}\slashed{l}))\gamma_5}{((l-k)^2 - M_V^2)(l^2 - M^2)}$$
$$- \int \frac{d^d l}{(2\pi)^d} \frac{2im\slashed{l} l^\rho \sigma_{\rho\tau} \gamma_5}{((\bar{p}-l)^2 - m^2)(l^2 - M^2)} \bigg).$$

For completeness, we shall give the relevant loop integrals with tensor structures in app. C. Using the coefficient functions defined there, the result for the first integral I_1 can be written as

$$I_1 = -ig_1 \tau^i (\gamma_\tau I_1^{(\gamma)} + k_\tau I_1^{(k)} + \bar{p}_\tau I_1^{(p)}) \gamma_5, \quad (2.33)$$

where the coefficients read

$$I_1^{(\gamma)} = 2m(M_V^2 - k^2)C_1 + mk^2(I_{MBV}^A + I_{MBV}^B)(M^2 + M_V^2 - k^2)$$
$$- mM^2(M^2 + M_V^2 - k^2)I_{MBV} + mM^2 I_{MB},$$
$$I_1^{(k)} = 2m^2(M_V^2 - k^2)(3C_2 + C_3 + 4C_4) + 4m^2 M^2(I_{MBV}^A + I_{MBV}^B)$$
$$- 2m^2(M_V^2 - k^2)(3I_{MBV}^A + I_{MBV}^B) + 2m^2(I_{MBV}^A - I_{MBV}^B)(M^2 + M_V^2 - k^2)$$
$$- 4m^2 M^2 I_{MBV} + t_{MV}^{(1)} + (M_V^2 - 2k^2)I_{MV}^{(1)} + (k^2 - M_V^2)I_{MV}$$
$$+ (M^2 + M_V^2 - k^2)(I_{MV} - I_{MV}^{(1)}),$$
$$I_1^{(p)} = 2m^2(M_V^2 - k^2)(3C_2 - C_3 - 2C_4) + 4m^2 M^2(I_{MBV}^A - I_{MBV}^B)$$
$$- 2m^2(I_{MBV}^A - I_{MBV}^B)(M^2 + M_V^2 - k^2) + 2t_{MV}^{(0)} - 2m^2 I_{MB}^{(1)}$$
$$+ (M^2 + M_V^2 - k^2)(I_{MV}^{(1)} - I_{MV}) - I_M.$$

What concerns the evaluation of I_2, we note that it is given by

$$I_2 = -ig_1 \tau^i (\gamma_\tau I_2^{(\gamma)} + k_\tau I_2^{(k)} + p_\tau I_2^{(p)}) \gamma_5, \qquad (2.34)$$

with

$$I_2^{(\gamma)} = I_1^{(\gamma)}, \quad I_2^{(k)} = I_1^{(k)}, \quad I_2^{(p)} = -I_1^{(p)}.$$

The sum of both graphs therefore gives

$$I_{1+2} = I_1 + I_2 = -ig_1 \tau^i (\gamma_\tau I_{1+2}^{(\gamma)} + k_\tau I_{1+2}^{(k)}) \gamma_5,$$

with

$$I_{1+2}^{(\gamma)} = 2I_1^{(\gamma)}, \quad I_{1+2}^{(k)} = 2I_1^{(k)} + I_1^{(p)}.$$

In the sum of the two graphs, the contribution proportional to $(\bar{p}+p)_\tau$ cancels, as was to be expected on general grounds (see the remarks following eq. (2.28)). We are now in a position to display the decomposition of the graphs as a linear combination of the scalar loop integrals worked out in the preceding sections:

$$I_{1+2}^{(\gamma)} = c_{MBV}^{(\gamma)} I_{MBV}^{IR} + c_{MB}^{(\gamma)} I_{MB}^{IR} + c_{MV}^{(\gamma)} I_{MV}^{IR}, \qquad (2.35)$$
$$I_{1+2}^{(k)} = c_{MBV}^{(k)} I_{MBV}^{IR} + c_{MB}^{(k)} I_{MB}^{IR} + c_{MV}^{(k)} I_{MV}^{IR} + c_M^{(k)} I_M. \qquad (2.36)$$

The expressions for the coefficients $c^{(\gamma)}$ read

$$c_{MBV}^{(\gamma)} = \frac{4m}{(d-2)k^2(k^2-4m^2)} \Big[m^2 M_V^6 + (k^2((d-5)m^2 + M^2) - 2m^2 M^2) M_V^4$$
$$- ((M^2 + m^2(2d-7))k^4 - 2(d-2)m^2 M^2 k^2 - m^2 M^4) M_V^2$$
$$+ (d-3)k^2 m^2 (k^2 - M^2)^2 \Big],$$
$$c_{MB}^{(\gamma)} = -\frac{2m}{(d-2)k^2(k^2-4m^2)} \Big[(M_V^2 + (d-3)k^2)((2m^2 - M^2)k^2 + 2m^2(M^2 - M_V^2)) \Big],$$
$$c_{MV}^{(\gamma)} = \frac{2m}{(d-2)(k^2-4m^2)} \Big[(M_V^2 + (d-3)k^2)(k^2 - M^2 - M_V^2) \Big],$$

and the coefficients $c^{(k)}$ are given by

$$c_{MBV}^{(k)} = \frac{2m^2}{(d-2)k^4(k^2-4m^2)} \Big[((d-2)k^2 - 4(d-1)m^2)M_V^6$$
$$- 2((d-2)k^4 + (dM^2 - 2(d+1)m^2)k^2 - 4(d-1)m^2 M^2)M_V^4$$
$$+ ((d-2)k^6 + 4(d-5)m^2 k^4 - 2(d-4)M^2 k^4 + ((d-2)k^2 - 4(d-1)m^2)M^4)M_V^2$$
$$- 4(d-3)k^2 m^2 (k^2 - M^2)^2 \Big],$$

$$c_{MB}^{(k)} = \frac{1}{(d-2)k^4(k^2-4m^2)} \Big[(4(d-3)m^2(2m^2 - M^2) - (d-2)(2m^2 + M^2)M_V^2)k^4$$
$$+ 2m^2((d-2)M_V^4 + (d-4)M^2 M_V^2 + 4m^2((d-3)M^2 + 2M_V^2))k^2$$
$$+ 8(d-1)m^4(M^2 - M_V^2)M_V^2 \Big],$$

$$c_{MV}^{(k)} = -\frac{4m^2}{(d-2)k^2(k^2-4m^2)} \Big[(M_V^2 + (d-3)k^2)(k^2 - M^2 - M_V^2) \Big],$$

$$c_M^{(k)} = \frac{M_V^2}{k^2}.$$

Here we used the abbreviations $k^4 \equiv (k^2)^2$ and $k^6 \equiv (k^2)^3$. In view of the denominators of the coefficients $c^{(\gamma,k)}$, which contain powers of k^2, it is advantageous to expand the scalar loop integrals in the small variable $\beta = \gamma\tilde{\beta}$ first. From eqs. (2.15,2.27), we find

$$I_{MV}^{IR} = \frac{I_M}{m^2(\alpha-\gamma)} + \beta \left(\frac{(d\gamma - (d-4)\alpha)I_M}{m^2 d(\alpha-\gamma)^3} \right) + \ldots$$

and

$$I_{MBV}^{IR} = \frac{I_{MB}^{IR}}{m^2(\alpha-\gamma)}$$
$$+ \beta \left(\frac{((d-1)(\gamma-\alpha)(\alpha-2) - 2(4\alpha-\alpha^2))m^2 I_{MB}^{IR} + ((d-3)\alpha - (d-1)\gamma)I_M}{2(d-1)m^4(\gamma-\alpha)^3} \right)$$
$$+ O(\beta^2).$$

Inserting the β-expansions of the scalar loop integrals in the expressions for $I_{1+2}^{(\gamma,k)}$ from eqs. (2.35) and (2.36), one observes that the poles in the variable $\beta \sim k^2$ cancel. In the final step, we must insert the expansion of the scalar loop integral I_{MB}^{IR} in the second small variable $\alpha \sim M^2$, which can directly be read off from eq. (2.8). The expression for I_M is given in eq. (2.2). Doing this, taking the limit $d \to 4$ and comparing to the decomposition of the matrix element in eq. (2.28), one finds the following $O(q^3)$-contribution of I_1 and I_2 to the axial form factor G_A:

$$G_A^{1+2} = -\frac{g_1}{3\pi} M^3 + O(q^4) = -\frac{2\sqrt{2} g_A G_V R_V}{3\pi M_V^2 F^2} M^3 + O(q^4). \tag{2.37}$$

There are no terms of lower order. This is in accord with the power counting for the two graphs, which predicts a chiral order of q^3 for this contribution to G_A. In order to compare this with the q^3-terms in G_A worked out in [64], one can proceed as follows. Looking at fig. 2.2, and imagining the vector meson lines shrinking to a point vertex (corresponding to a limit where the mass M_V tends to infinity, with $G_V R_V/M_V$ fixed), it is intuitively clear that the result corresponds to a pion-nucleon loop graph with an $O(q^2)$ contact term replacing the vector meson line. Such contributions are parametrized by the two LECs c_3 and c_4 in [64], see eq. (2.29). In fact, identifying

$$\frac{2\sqrt{2} G_V R_V}{M_V^2} = -c_4, \qquad (2.38)$$

one reproduces exactly the corresponding terms in the representation based on the pure pion-nucleon theory, cf. eq. (2.29). That eq. (2.38) is a good guess can be seen like that: Comparing the ρN-coupling used here, namely, the term proportional to R_V in eq. (2.32), to a more conventional one using a vector field representation for the rho field,

$$\mathcal{L}_{NV} = \frac{1}{2} g_{\rho NN} \bar{\psi} \left(\gamma^\mu \rho_\mu \cdot \tau - \frac{\kappa_\rho}{2m} \sigma^{\mu\nu} \partial_\nu \rho_\mu \cdot \tau \right) \psi, \qquad (2.39)$$

one deduces

$$g_{\rho NN} \kappa_\rho = -\frac{4\sqrt{2} m R_V}{M_V}.$$

Using this in eq. (2.38), we get

$$c_4 = \frac{g_{\rho NN} \kappa_\rho G_V}{2m M_V} = \frac{\kappa_\rho}{4m}.$$

In the last step, we have assumed a universal rho coupling, $M_V G_V \equiv F^2 g_{\rho\pi\pi} = F^2 g_{\rho NN}$ as well as the KSFR relation $M_V^2 = 2F^2 g_{\rho NN}^2$ [66] (see also the recent discussion in the framework of effective field theory in ref. [67]). This agrees with the rho-contribution to c_4 found in [43]. Furthermore, there is no rho contribution to the LEC c_3 according to this work.

The result of eq. (2.38) is not surprising for itself, but the agreement of our findings with previous resonance saturation analyses demonstrates one very important thing, namely, that the variants of the infrared regularization scheme derived in the previous sections are consistent with the standard case of infrared regularization in the pion-nucleon sector used in [64].

As a side remark, we note that the leading order result from the triangle graphs shows no t-dependence and therefore gives no contribution to the axial radius. However, the leading contribution to the axial radius can also be related to meson resonances, namely, to a tree-level exchange of an axial-vector meson. The pertinent calculation can be found in [68], where the axial vector meson-couplings to the pions and nucleons are fitted to experimental data for $G_A(t)$ (for an earlier study based on chiral Lagrangians, see [69]). Equivalently, it can be parametrized by a certain LEC, named d_{22} in [64] (see eq. (2.29)). As already mentioned at the beginning of sec. 2.5, the difference between the two approaches just amounts to a resummation of higher order terms. Compared to the leading order term, the t-dependent part derived from the triangle graphs is suppressed by factors of the small variable α, which is a reflection of the fact that the infrared regularized loop integrals preserve the chiral power counting.

2.6 Summary

In this chapter we have presented an extension of the infrared regularization scheme that allows for an inclusion of explicit (vector and axial-vector) meson resonances in the single-nucleon sector of ChPT. For the processes we have considered here, the meson resonances do not appear as external particles, and the corresponding power counting rules for the internal resonance lines are set up such that the resonance four-momentum is considered to be small compared to its mass. The infrared regularization scheme extracts the part of the one-loop graphs to which this power counting scheme applies (for any value of the dimension parameter d used in dimensional regularization), while the remaining parts of the loop graphs will in general violate the power counting requirements, but can be absorbed in a renormalization of the local terms of the effective Lagrangian.

After a short review of the infrared regularization procedure used for the pion-nucleon and the vector meson-pion system in sec. 2.2 and 2.3, respectively, we have combined the analyses of these sections in sec. 2.4. There, we consider the simplest example of a Feynman graph where nucleons, pions as well as (vector) meson resonances show up. It is shown how to extract the infrared singular part of such a graph, and the power counting requirements are verified. It should be clear from this example how the infrared singular parts of more complicated one-loop graphs (with more nucleon, pion and resonance lines) can be worked out (some remarks on the general case can be found at the beginning of sec. 2.4, and in sec. 6 of [38]). Finally, in sec. 2.5, we have applied the extended scheme to compute a vector meson induced loop contribution to the axial form factor of the nucleon, and demonstrate that the result agrees with the result for $G_A(t)$ given by Schweizer [64] in combination with the resonance saturation analysis for the pion-nucleon LECs in [43]. There are, of course, many other possibilities for applications of the scheme developed here. Finally, we add the remark that a suggestion for an extension of infrared regularization to the multiloop case was made in ref. [70].

Chapter 3

Quark mass dependence of the mass of the Roper(1440)

3.1 Introduction

In the previous chapter, we have developed an extension of the standard infrared regularization scheme [38] to the case of explicitly included vector mesons in baryon ChPT. This extension was guided by what we called the operational definition of infrared regularization in sec. 2.2, namely, the extraction of those parts of the loop integrals which are proportional to d-dependent powers of the small expansion parameters of ChPT. As this principle is fairly general, it might be interesting to look for other situations where it can be applied. In this chapter, we shall consider the explicit inclusion of a baryonic resonance in BChPT, namely, the inclusion of the so-called Roper resonance, the mass of which is about ~ 1440 MeV [15]. The Roper $N^*(1440)$ is the first even-parity excited state of the nucleon, and its examination is quite intriguing—it is lighter than the first odd-parity nucleon excitation, the $S_{11}(1535)$, and also has a significant branching ratio into a nucleon with two pions. Recent lattice studies, see e.g. [71–77], have not offered a clear picture about the nucleon resonance spectrum. The findings of ref. [72] indicate a rapid cross over of the first positive and negative parity excited nucleon states close to the chiral limit. No such level switching is e.g. found in [74], possibly related to the fact that the simulations were performed at quark masses too far away from the chiral regime. One has to remember that lattice QCD operates at unphysical quark (pion) masses and thus a chiral extrapolation is needed to connect these data to the physical world, as already mentioned in the introduction to this thesis, sec. 1.1. It should also be noted that so far very simple chiral extrapolation functions have been employed in most approaches, e.g., a linear extrapolation in the quark masses, thus $\sim M_\pi^2$ (with M_π the pion mass), was applied in [77]. It is therefore important to provide the lattice practitioners with improved chiral extrapolation functions. This is the aim of the present investigation. We consider the Roper mass (the real part of the Roper self–energy) to one-loop in baryon chiral perturbation theory, employing a novel, tailor-made extension of the standard infrared regularization scheme. This enables us to study the corresponding pion mass dependence, parametrized by the various low-energy (coupling) constants that appear in the expression, in a model-independent way. We refrain from analyzing the existing lattice data—our results apply to full QCD and not to one of the various approximations to QCD employed in the lattice studies.

This chapter is organized as follows. In sec. 3.2, we display the effective chiral Lagrangian underlying our calculation. The chiral corrections to the Roper mass are calculated in sec. 3.3, where we also explain the relevant extension of infrared regularization. Sec. 3.4 contains our results and the discussion thereof.

3.2 Effective Lagrangian

It is our aim to calculate chiral corrections to the Roper mass up to one-loop order. Since the Roper is the first even-parity excited state of the nucleon, the construction of the chiral $SU(2)$ effective Lagrangian follows standard procedures, see e.g. [78]. The effective Lagrangian relevant for our calculation is (see also ref. [79])

$$\mathcal{L} = \mathcal{L}_0 + \mathcal{L}_R + \mathcal{L}_{NR} ~, \tag{3.1}$$

with the free part

$$\mathcal{L}_0 = i\bar{N}\gamma_\mu D^\mu N - m_N \bar{N}N + i\bar{R}\gamma_\mu D^\mu R - m_R \bar{R}R ~, \tag{3.2}$$

where N, R are nucleon and Roper fields, respectively, and m_N, m_R the corresponding baryon masses in the chiral limit. D_μ is the chiral covariant derivative, for our purpose we can set $D_\mu = \partial_\mu$, see sec. 1.3 or e.g. [78] for definitions. The pion-Roper coupling is given to leading chiral order by

$$\mathcal{L}_R^{(1)} = \frac{1}{2} g_R \bar{R}\gamma_\mu \gamma_5 u^\mu R \tag{3.3}$$

with an unknown coupling g_R, and the superscript denotes the chiral order. The pion fields are collected in $u_\mu = -\partial_\mu \pi/F + \mathcal{O}(\pi^3)$, where F is the pion decay constant in the chiral limit (see also sec. 1.2). At next-to-leading order, the relevant terms in $\mathcal{L}_R$ are (we work in the isospin limit $m_u = m_d$ and neglect electromagnetism)

$$\mathcal{L}_R^{(2)} = c_1^* \langle \chi_+ \rangle \bar{R}R - \frac{c_2^*}{8m_R^2}\bar{R}(\langle u_\mu u_\nu \rangle \{D^\mu, D^\nu\} + \text{h.c.})R + \frac{c_3^*}{2}\langle u_\mu u^\mu \rangle \bar{R}R ~, \tag{3.4}$$

where

$$\chi_+ = 2B(u^\dagger \mathcal{M} u^\dagger + u\mathcal{M}u) \tag{3.5}$$

describes explicit chiral symmetry breaking via the quark mass matrix $\mathcal{M} = \text{diag}\,(m_u, m_d, m_s)$. For a complete one loop calculation we also need the fourth order effective Lagrangian, more precisely, the term

$$\mathcal{L}_R^{(4)} = -\frac{e_1^*}{16}\langle \chi_+ \rangle^2 \bar{R}R ~. \tag{3.6}$$

The interaction piece between nucleons and the Roper reads

$$\mathcal{L}_{NR}^{(1)} = \frac{1}{2}g_{NR}\bar{R}\gamma_\mu \gamma_5 u^\mu N + \text{h.c.} ~. \tag{3.7}$$

The coupling g_{NR} can be determined from the strong decays of the resonance R, its actual value is given below. In principle a term of the form

$$i\lambda_1 \bar{R}\gamma_\mu D^\mu N - \lambda_2 \bar{R}N + \text{h.c.} \tag{3.8}$$

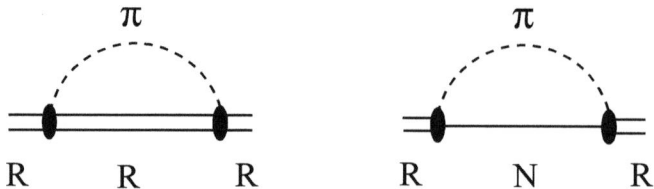

Figure 3.1: One-loop self-energy graphs of the Roper (R) with intermediate Roper-pion (π) and nucleon (N)-pion states, respectively.

is possible, but applying the equations of motion removes the first term (and its hermitian conjugate) such that we are left with the terms $\bar{R}N$ and $\bar{N}R$. These terms induce mixing between the nucleon and Roper fields, but diagonalization of the N-R mass matrix does not lead to new operator structures and its effect can be completely absorbed into the couplings already present in the Lagrangian. We can thus safely work with the Lagrangian in eq. (3.1). A complete one-loop calculation involves tree graphs with insertion of chiral dimension two and four and one-loop graphs with at most one insertion from $\mathcal{L}_R^{(2)}$.

3.3 Chiral corrections to the Roper mass

We are now in the position to work out the various contributions to the Roper mass. The loop diagrams are evaluated making use of (an extension of) the infrared regularization (IR) method [38]. As we will see, the IR scheme is suited for the study of systems with one light mass scale M_π and two heavy mass scales m_N, m_R with $m_N^2 \ll m_R^2$. In the real world, we have $m_R^2/m_N^2 \simeq 2.4$, so that this condition is approximately fulfilled.

1. Tree level: Only the c_1^* and e_1^* terms contribute to the self-energy at tree level

$$\Sigma_R^{\text{tree}} = -4c_1^* M_\pi^2 + e_1^* M_\pi^4 \ . \tag{3.9}$$

These terms could be absorbed into m_R, but since we are interested in the explicit dependence on the pion (quark) mass, we must keep them. The first term is the leading order contribution to the nucleon-Roper σ-term.

2. Pion-nucleon loop: This is the only new structure compared e.g. to the calculation of the nucleon self-energy, see fig. 3.1. Here we extend the method of chapter 2 and ref. [51], developed for IR with vector mesons. Consider first the fundamental scalar integral at one-loop order in d dimensions with an intermediate pion-nucleon pair and external momentum p

$$I_{\pi N}(p^2) = \int \frac{d^d l}{(2\pi)^d} \frac{i}{[l^2 - M_\pi^2 + i\epsilon][(p+l)^2 - m_N^2 + i\epsilon]} \ . \tag{3.10}$$

Employing standard Feynman parameterization leads to

$$I_{\pi N}(p^2) = -\frac{\Gamma(2-\frac{d}{2})}{(4\pi)^{d/2}} m_N^{d-4} \int_0^1 dz \ (\beta[z-x_+][z-x_-])^{\frac{d}{2}-2} \tag{3.11}$$

in terms of the parameters

$$x_\pm = \frac{\alpha+\beta-1}{2\beta}\left(1\pm\sqrt{1-\frac{4\alpha\beta}{(\alpha+\beta-1)^2}}\right), \quad \alpha = \frac{M_\pi^2}{m_N^2}, \quad \beta = \frac{p^2}{m_N^2}. \quad (3.12)$$

In the following, we will restrict ourselves to the kinematical region $p^2 \gg (m_N + M_\pi)^2$ which is equivalent to

$$\frac{4\alpha\beta}{(\alpha+\beta-1)^2} \ll 1. \quad (3.13)$$

This implies that in the chiral limit $\alpha \to 0$ the mass difference $p^2 - m_N^2$ remains finite and does not tend towards zero as in the standard case of IR. This constraint is clearly satisfied for values $p^2 \approx m_R^2$ close to the Roper mass. It is also consistent with resonance decoupling in the chiral limit [80]. The nucleon propagator is thus counted as zeroth chiral order for momenta $p^2 \approx m_R^2$. Setting $\beta = m_R^2/m_N^2$ one obtains the small parameter

$$\frac{4\alpha\beta}{(\alpha+\beta-1)^2} \approx \frac{1}{9}, \quad (3.14)$$

which indicates a fast convergence of the expansion of the loop integral eq. (3.10) in powers of the pion mass. Expansion of $x_\pm$ in α leads to

$$x_+ = \frac{\beta-1}{\beta} - \frac{\alpha}{\beta(\beta-1)} - \frac{\alpha^2}{(\beta-1)^3} + \mathcal{O}(\alpha^3),$$

$$x_- = \frac{\alpha}{\beta-1} + \frac{\alpha^2}{(\beta-1)^3} + \mathcal{O}(\alpha^3), \quad (3.15)$$

where $x_- = \mathcal{O}(\alpha)$ and $x_+ = \mathcal{O}(1)$. We now divide the parameter integral of eq. (3.11) into three parts

$$I_{\pi N} = -\frac{\Gamma(2-\frac{d}{2})}{(4\pi)^{d/2}} m_N^{d-4}\left(I_{\pi N}^{(1)} + I_{\pi N}^{(2)} + I_{\pi N}^{(3)}\right), \quad (3.16)$$

with

$$I_{\pi N}^{(1)}(p^2) = \int_0^{x_-} dz\,(\beta[z-x_+][z-x_-])^{\frac{d}{2}-2},$$

$$I_{\pi N}^{(2)}(p^2) = \int_{x_-}^{x_+} dz\,(\beta[z-x_+][z-x_-])^{\frac{d}{2}-2},$$

$$I_{\pi N}^{(3)}(p^2) = \int_{x_+}^1 dz\,(\beta[z-x_+][z-x_-])^{\frac{d}{2}-2}. \quad (3.17)$$

Note that $0 < x_- < x_+ < 1$. The first integral $I_{\pi N}^{(1)}$ can be rewritten as

$$I_{\pi N}^{(1)}(p^2) = (x_-)^{d/2-1} \int_0^1 dy\,(\beta[x_+ - x_-(1-y)])^{\frac{d}{2}-2} y^{\frac{d}{2}-2}. \quad (3.18)$$

Expansion of the integrand in powers of $x_- \sim \mathcal{O}(\alpha)$ and interchanging summation with integration leads to

$$I_{\pi N}^{(1)}(p^2) = \frac{2}{d-2}(x_-)^{d/2-1}(\beta x_+)^{d/2-2} + \mathcal{O}(\alpha^{d/2}), \quad (3.19)$$

where the leading term is of order $\mathcal{O}(\alpha^{d/2-1})$ and thus conserves power counting. The integral $I_{\pi N}^{(1)}$ contains only fractional powers of α and contributes to the infrared singular part. The remaining two integrals, on the other hand, are regular in α. For $I_{\pi N}^{(2)}$ one has

$$I_{\pi N}^{(2)}(p^2) = (-\beta)^{d/2-2} \int_{x_-}^{x_+} dz \; ([x_+ - z][z - x_-])^{\frac{d}{2}-2} \;, \qquad (3.20)$$

where β has a small positive imaginary piece. Integration yields

$$I_{\pi N}^{(2)}(p^2) = (-\beta)^{d/2-2} (x_+ - x_-)^{d-3} \frac{(\Gamma(\frac{d}{2}-1))^2}{\Gamma(d-2)} \;. \qquad (3.21)$$

Obviously, this expression is expandable in powers of x_-. The integral $I_{\pi N}^{(2)}$ is complex and does not conserve power counting. It contributes only to the regular part and will be omitted for our purposes. More precisely, the imaginary part does not contribute to the resonance mass, while the real part can be absorbed into the couplings of the effective Lagrangian at the on-shell momentum $p^2 = m_R^2$. In the third integral,

$$I_{\pi N}^{(3)}(p^2) = \int_{x_+}^{1} dz \; (\beta[z - x_+][z - x_-])^{\frac{d}{2}-2} \;, \qquad (3.22)$$

one can expand the integrand directly in powers of x_-. The expansion coefficients of this series are integrals of the type ($r \in \mathbb{R}$)

$$\int_{x_+}^{1} dz \; (z - x_+)^{\frac{d}{2}-2} z^r = (1 - x_+)^{d/2-1} \int_0^1 dw \; (1-w)^{\frac{d}{2}-2} (1 + w(x_+ - 1))^r \;. \qquad (3.23)$$

Since $1 - x_+$ remains finite in the chiral limit these integrals also contribute only to the regular part and can be absorbed into the couplings of the Lagrangian at momentum $p^2 = m_R^2$. Therefore, the infrared singular part which stems from small values of the Feynman parameter z is entirely contained in $I_{\pi N}^{(1)}$ and we can restrict ourselves to the integral

$$\begin{aligned} I_{\pi N}^{IR} &\equiv -\frac{\Gamma(2 - \frac{d}{2})}{(4\pi)^{d/2}} m_N^{d-4} \int_0^{x_-} dz \; (\beta[x_+ - z][x_- - z])^{\frac{d}{2}-2} \\ &= -\frac{m_N^{d-4}}{16\pi^2} \left\{ x_- \left(\frac{2}{4-d} + \ln 4\pi - \gamma_E + 1 \right) - \int_0^{x_-} dz \; \ln \left(\beta[x_+ - z][x_- - z] \right) \right\} \;. \end{aligned} \qquad (3.24)$$

Expansion in α leads to

$$I_{\pi N}^{IR} = \left(2L + \frac{1}{16\pi^2} \ln \left(\frac{M_\pi^2}{m_R^2} \right) \right) \left(\frac{\alpha}{\beta - 1} + \frac{\alpha^2}{(\beta - 1)^3} \right) - \frac{1}{32\pi^2} \frac{\alpha^2 \beta}{(\beta - 1)^3} + \mathcal{O}(M_\pi^6) \;, \qquad (3.25)$$

with

$$L = \frac{m_R^{d-4}}{16\pi^2} \left\{ \frac{1}{d-4} - \frac{1}{2} [\ln 4\pi - \gamma_E + 1] \right\} \;, \qquad (3.26)$$

and γ_E is the Euler-Mascheroni constant. We have chosen the regularization scale to be m_R.

In fact, the same result for the infrared part can be obtained by expanding the baryon propagator in the loop integral (a method first used in ref. [39])

$$\int \frac{d^d l}{(2\pi)^d} \frac{i}{[l^2 - M_\pi^2] [p^2 + 2p \cdot l + l^2 - m_N^2]} \,. \tag{3.27}$$

Counting the loop momentum as $l \sim \mathcal{O}(M_\pi)$ and taking $p^2 \gg (m_N + M_\pi)^2$, one obtains

$$\frac{1}{p^2 - m_N^2} \int \frac{d^d l}{(2\pi)^d} \frac{i}{l^2 - M_\pi^2} \left(1 - \frac{2p \cdot l + l^2}{p^2 - m_N^2} + \frac{(2p \cdot l + l^2)^2}{(p^2 - m_N^2)^2} + \ldots \right) , \tag{3.28}$$

which reproduces the result of eq. (3.25) after interchanging summation and integration. The observation that the procedure of ref. [39] reproduces the infrared singular parts of loop integrals was also made in the previous chapter.

After investigating the scalar loop integral one readily obtains the infrared singular part of the full one-loop self-energy diagram with an intermediate pion-nucleon pair (see fig. 3.1)

$$\begin{aligned} (\Sigma_N(\slashed{p}))_{IR} &= i \frac{3g_{NR}^2}{4F^2} \int_{IR} \frac{d^d l}{(2\pi)^d} \frac{\slashed{l}(\slashed{p} + \slashed{l} - m_N)\slashed{l}}{[l^2 - M_\pi^2] [(p + l)^2 - m_N^2]} \\ &= -\frac{3g_{NR}^2}{4F^2} \left(\slashed{p} \left[\frac{M_\pi^4}{32\pi^2(p^2 - m_N^2)} \ln\left(\frac{M_\pi^2}{m_R^2}\right) + \frac{M_\pi^4}{64\pi^2(p^2 - m_N^2)} \right] \right. \\ &\quad \left. + \frac{m_N M_\pi^4}{16\pi^2(p^2 - m_N^2)} \ln\left(\frac{M_\pi^2}{m_R^2}\right) \right) , \end{aligned} \tag{3.29}$$

where we have only displayed the finite pieces. The term proportional to L has been absorbed into the counter terms. Evaluating the integral at $\slashed{p} = m_R$ yields

$$(\Sigma_N(m_R))_{IR} = -\frac{3g_{NR}^2}{256\pi^2 F^2 (m_R^2 - m_N^2)} M_\pi^4 \left[(2m_R + 4m_N) \ln\left(\frac{M_\pi^2}{m_R^2}\right) + m_R \right] . \tag{3.30}$$

Note that this expression preserves both power counting and chiral symmetry.

3. Pion-Roper loop: This corresponds to the standard IR case and is immediately obtained from the result in [38] by replacing m_N by m_R (see fig. 3.1),

$$(\Sigma_R(m_R))_{IR} = -\frac{3g_R^2}{32\pi F^2} M_\pi^3 \left[1 + \frac{M_\pi}{2\pi m_R} + \frac{M_\pi}{2\pi m_R} \ln\left(\frac{M_\pi^2}{m_R^2}\right) \right] + \mathcal{O}(M_\pi^5) . \tag{3.31}$$

Again, power counting and chiral symmetry are maintained for the IR singular part of this diagram.

4. Tadpoles: The tadpoles with vertices from $\mathcal{L}_R^{(2)}$ yield

$$\Sigma_R^{\text{tad}} = \left(6c_1^* - \frac{3}{4}c_2^* - 3c_3^* \right) \frac{M_\pi^4}{16\pi^2 F^2} \ln\left(\frac{M_\pi^2}{m_R^2}\right) + \frac{3}{128\pi^2 F^2} c_2^* M_\pi^4 . \tag{3.32}$$

Again, this result agrees with the one for the nucleon by proper substitution of the LECs and baryon masses.

5. Total Roper self-energy: Putting all these pieces together, we obtain the following one-loop correction to the Roper mass

$$\delta m_R^{(1-\text{loop})} = \left(\Sigma_N(m_R)\right)_{IR} + \left(\Sigma_R(m_R)\right)_{IR} + \Sigma_R^{\text{tad}} + \Sigma_R^{\text{tree}} \qquad (3.33)$$

in terms of the renormalized couplings c_1^*, e_1^* and the renormalized mass m_R for which we use the same notation.

We have not explicitly considered loops with a $\Delta(1232)$-pion pair[1]. If one treats the Δ on the same footing as the nucleon field, the contribution will be of the type in eq. (3.30) and amounts to a renormalization of g_{NR} and e_1^*. However, one must keep in mind that the convergence of the corresponding chiral series is not as good as for the nucleon due to the smaller mass difference $m_R^2 - m_\Delta^2$. If, on the other hand, the Delta mass is considered to be of the same size as the Roper mass, the loop contribution will be similar to the result in eq. (3.31) and leads to a modification of g_R and the couplings c_i^*, e_1^*. In both scenarios, the effects of the $\Delta \pi$ loop can be absorbed into a redefinition of the couplings. Since their values are not fixed, we will vary them within certain regions, see sec. 3.4, such that the inclusion of the Δ resonance will not alter any of our conclusions. For a treatment of the Δ in infrared regularization see, e.g., refs. [47–49]. We also stress that the explicit inclusion of the Δ can lead to a complicated three-small-scales problem (the pion mass and—if considered small—the Roper-Delta and the Delta-nucleon splitting) that requires theoretical tools that have not yet been developed for baryon chiral perturbation theory.

3.4 Quark mass dependence of the Roper mass

Before analyzing the pion mass dependence of the Roper mass, we must collect information on the couplings g_R, g_{NR} and the LECs c_i^* ($i = 1, 2, 3$), and e_1^*. One obtains $g_{NR} = 0.3\ldots0.4$ from a fit to the branching ratio of the Roper into one pion and a nucleon which is in agreement with the relation $g_{NR} = \sqrt{R} g_A/2$, where g_A is the axial-vector coupling of the nucleon and $\sqrt{R} = 0.53 \pm 0.04$ (for details, see ref. [82]). For g_R the naive quark model predicts $g_R = g_A$, and we set here $g_R = 1.0$ so that g_A and g_R are roughly of the same size, see also [83].

To leading order in the chiral expansion, the LEC c_1^* measures the σ-term in the Roper state and it is thus bounded from above by the value of the pion-nucleon σ-term. This means $|c_1^*| \lesssim 1\,\text{GeV}^{-1}$. More realistically, a natural value for c_1^* would be around $-0.5\,\text{GeV}^{-1}$ because σ-terms are expected to become smaller with the resonance excitation energy (see also the related discussion on the $\pi\Delta$ σ-term in refs. [48,49,84,85]). The sign of c_1^* should be negative since the quark masses contribute positively to the hadron masses. The nucleon LECs c_2 and c_3 are much enhanced compared to the natural values $|c_i| \lesssim 1\,\text{GeV}^{-1}$ because of the nearby and strongly coupled delta resonance [43]. This is not expected to be the case for the Roper resonance. Consequently, the LECs $c_{2,3}^*$ can be bounded conservatively by $\pm 1\,\text{GeV}^{-1}$. The pion decay constant in the chiral limit is taken to be $F = 87$ MeV [20, 86].

In fig. 3.2 an estimated range for the pion mass dependence of the Roper mass is presented by taking the extreme values for $c_{2,3}^*$ and e_1^*, while keeping $c_1^* = -0.5\,\text{GeV}^{-1}$, $g_{NR} = 0.35$, $g_R = 1$

---

[1]Note that in the Jülich coupled-channels approach, the Roper is dynamically generated with an important $\pi\Delta$ component besides the σN one [81].

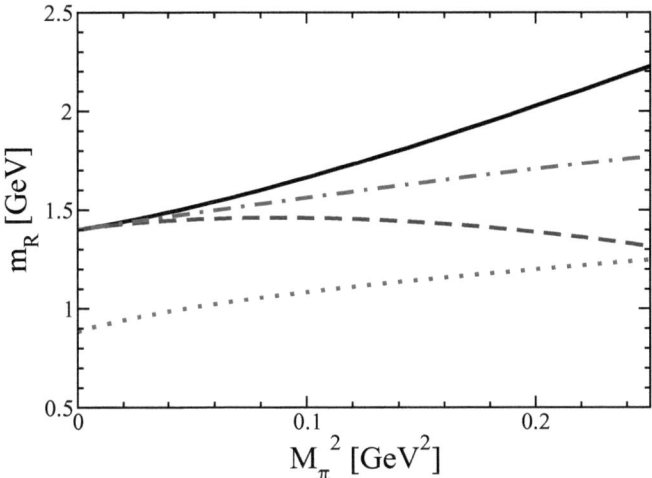

Figure 3.2: Quark mass dependence of the Roper mass for different parameter sets $c_1^* = -0.5, c_{2,3}^*, e_1^*$. The c_i are in units of GeV^{-1} and e_1 is given in GeV^{-3}. The axial couplings are $g_R = 1.0, g_{NR} = 0.35$. The solid curve corresponds to $c_2^* = 1.0, c_3^* = 1.0, e_1^* = 0.5$, the dashed one to $c_2^* = -1.0, c_3^* = -1.0, e_1^* = -0.5$ and the dot–dashed one to $c_2^* = c_3^* = e_1^* = 0$. The dotted curve represents the quark mass dependence of the nucleon, see ref. [87]. The values of the corresponding LECs are: $c_1 = -0.9, c_2 = 3.2, c_3 = -3.45, e_1 = -1.0$.

fixed. The masses of the baryons in the chiral limit are taken to be $m_N = 0.885$ GeV [87] and $m_R = 1.4$ GeV, respectively. The dash-dotted curve is obtained by setting the couplings $c_{2,3}^*, e_1^*$ all to zero, and exhibits up to an offset a similar quark mass dependence as the nucleon result (dotted curve, taken from ref. [87]). It should be emphasized, however, that the one-loop formula cannot be trusted for pion masses much beyond 350 MeV. For similar results for the nucleon mass, see also ref. [88].

In the numerical calculation we have employed the pion decay constant in the chiral limit, $F = 87$ MeV. However, we could have equally well used the physical pion decay constant, $F_\pi = 92.4$ MeV, as the difference in the chiral expansions appears either at chiral order $\mathcal{O}(M_\pi^6)$ in eqs. (3.30, 3.32) or at order $\mathcal{O}(M_\pi^5)$ for the Roper-pion loop which is beyond the accuracy of the present investigation. The numerical results for these contributions would change by the amount of $F^2/F_\pi^2 \approx 0.89$ and do not lead to significant changes in the results. Stated differently, the replacement of F by F_π in these formulae induces a correction due to the quark mass expansion of F_π

$$F_\pi = F\left(1 + \frac{M_\pi^2}{16\pi^2 F^2}\bar{l}_4 + \mathcal{O}(M_\pi^4)\right) \qquad (3.34)$$

with $\bar{l}_4 = 4.33$. The modifications at leading order are then

$$-\frac{3g_{NR}^2}{2048\pi^4 F_\pi^4 (m_R^2 - m_N^2)}\,\bar{l}_4\,M_\pi^6\left[(2m_R + 4m_N)\ln\left(\frac{M_\pi^2}{m_R^2}\right) + m_R\right] \qquad (3.35)$$

for the nucleon-pion loop,

$$-\frac{3g_R^2}{256\pi^3 F_\pi^4}\, \bar{l}_4\, M_\pi^5 \qquad (3.36)$$

for the Roper-pion loop, and

$$\left(6c_1^* - \frac{3}{4}c_2^* - 3c_3^*\right)\, \bar{l}_4\, \frac{M_\pi^6}{128\pi^4 F_\pi^4} \ln\left(\frac{M_\pi^2}{m_R^2}\right) + \frac{3}{1024\pi^4 F_\pi^4}c_2^*\, \bar{l}_4\, M_\pi^6 \, . \qquad (3.37)$$

for the tadpoles. These corrections at higher chiral orders are indeed small and can be safely neglected for small pion masses. In fact, the variations induced by these corrections are within the band for $m_R(M_\pi^2)$ given in fig. 3.2.

To summarize, we have calculated in this chapter the chiral corrections to the Roper mass to one-loop order. The approach is based on an extension of infrared regularization which allows for the unambiguous isolation of the infrared singular part of the loop integrals stemming from the pion poles of the pertaining integrands. At the same time, chiral symmetry is preserved and a chiral counting scheme emerges. The considered Feynman diagrams contain two different heavy mass scales m_N, m_R which we consider to satisfy the relation $m_N^2 \ll m_R^2$. The utilized formalism is in general suited to study systems with two heavy mass scales in addition to a light mass scale. In this sense, it can be applied to other resonances as well, such as the $S_{11}(1535)$. In this case, however, an SU(3) calculation is necessary due to the important ηN decay channel.

Chapter 4

Gauge invariance in unitarized ChPT

4.1 Introduction

In the foregoing two chapters, we have examined how the low-energy effective field theory of QCD, which is usually set up as a theory where only the Goldstone bosons (and possibly the ground-state baryons) are treated as dynamical degrees of freedom, is affected by the appearance of an additional mass scale, given by the mass of an explicitly included resonance field. We have shown that, in various cases, the method of infrared regularization can suitably be generalized to establish a well-defined power counting scheme also for such extended versions of the original effective field theory, at least at the one-loop level.

In the remaining part of this thesis, we will focus on another (non-perturbative) class of extensions of the standard version of ChPT, which was already introduced in sec. 1.5 as unitarized chiral perturbation theory (UChPT). Over the last years, considerable effort has been undertaken to combine the effective chiral Lagrangian approach with such non-perturbative methods, both in the purely mesonic sector [55] and in the meson-baryon sector [56–58]. Recall that the systematic loop expansion of standard ChPT involves a characteristic scale $\Lambda_\chi \simeq 4\pi F_\pi \approx 1.2$ GeV at which the chiral series is expected to break down. The limitation to very low-energy processes is even enhanced in the vicinity of resonances. The appearance of resonances in certain channels constitutes a major problem for the loop expansion of ChPT, since a resonance cannot be reproduced at any given order of the chiral series. Nevertheless, at low energies the contribution from such resonances is encoded in the numerical values of certain low-energy constants, as explained in chapter 1.

The combination with the non-perturbative schemes of UChPT has made it possible to go to energies beyond Λ_χ and to generate resonances dynamically (giving up, however, certain aspects of the rigorous framework constituted by ChPT). Two prominent examples in the baryonic sector are the $\Lambda(1405)$ and the $S_{11}(1535)$. The first one is an s-wave resonance just below the K^-p threshold and dominates the interaction of the $\bar{K}N$ system, while the $S_{11}(1535)$ was identified in [56] as a quasi-bound $K\Lambda$-$K\Sigma$ state.

Such chiral unitary approaches have been extended to photo- and electroproduction processes of mesons on baryons, see e.g. [89–91]. In these coupled channel models the initial photon scatters with the incoming baryon into a meson-baryon pair which in turn rescatters (elastically or inelastically) an arbitrary number of times. The two-body final state interactions are taken into account in a coupled-channels Bethe-Salpeter equation (BSE) or – in the non-relativistic framework – Lippmann-Schwinger equation. The coupling of the incoming photon to other possible vertices is omitted, and although these approaches appear to describe the available data well, the issue of gauge invariance is not discussed in these chiral unitary approaches. On the other hand, a method to obtain conservation of the electromagnetic current of a two-nucleon system is presented in [92] and extended to a resonance model for pion photoproduction in [93]. Alternatively, the so-called "gauging of equations" method has been developed in [94, 95] to incorporate an external electromagnetic field in the integral equation of a few-body system in a gauge invariant fashion. Further investigations of gauge invariance in pion photoproduction within πN models can be found in [96, 97]. Gauge invariance is also of interest in coupled-channels approaches in the mesonic sector, e.g in radiative ϕ, ρ decays [98, 99] and in the anomalous decays $\eta, \eta' \to \gamma\gamma, \pi^+\pi^-\gamma$ [100, 101]. For related recent work, see also [102].

The purpose of the present chapter is to illustrate how gauge invariance can be maintained when an external photon couples to a two-particle state described by the BSE within the chiral effective framework. The method developed here will be applied in the next chapter to construct an amplitude for meson photo-and electroproduction which fulfills the requirements of exact two-body unitarity and gauge invariance. We will start in the next section by first outlining the procedure for the simple scalar field theory we have already introduced in sec. 1.5. With the insights gained from this example we can then address gauge invariance in meson-baryon scattering processes with chiral effective Lagrangians. In sec. 4.3 we discuss the case with the interaction kernel of the BSE derived from the leading order Lagrangian – the Weinberg-Tomozawa term. The extension to driving terms from higher chiral orders is presented in sec. 4.4. In section 4.5 it is shown that the corresponding amplitudes also satisfy unitarity constraints. Our conclusions and outlook are presented in sec. 4.6.

4.2 Complex scalar fields

In this section, we will demonstrate in a simple field theory with complex scalar fields that the coupling of an external photon to a two-body state, which corresponds to the solution of the BSE, is gauge invariant. To this end, we consider the (normal-ordered) Lagrangian for complex fields ϕ and ψ

$$\mathcal{L} = \partial_\mu \phi^* \partial^\mu \phi - m^2 \phi^* \phi + \partial_\mu \psi^* \partial^\mu \psi - M^2 \psi^* \psi - g(\phi^* \phi)(\psi^* \psi) , \tag{4.1}$$

with masses m and M, respectively, which was already used as a toy-model in sec. 1.5. For the purpose of a self-contained presentation, we shall repeat some of the features of this toy-model in this section.

For small values of the coupling constant g, the scattering process $\phi(p_1)\psi(p_2) \to \phi(p_3)\psi(p_4)$ may be calculated perturbatively. For general values of g, however, and if one is interested in bound states, one must resort to non-perturbative techniques such as the BSE. The Bethe-Salpeter equation for the scattering matrix T of the two-particle scattering process, eq. (1.50), reduces to a simple algebraic equation in the present case, namely

$$T(s) = g + T(s)G(s)g , \tag{4.2}$$

where $s = p^2 = (p_1+p_2)^2 = (p_3+p_4)^2$ and G is the scalar loop integral from eq. (1.46),

$$G(p^2) = i \int_l \Delta_\phi(p-l)\Delta_\psi(l) \tag{4.3}$$

utilizing the short-hand notation

$$\int_l = \int \frac{d^4l}{(2\pi)^4} \qquad (4.4)$$

and the propagators

$$i\Delta_\phi(l) = \frac{i}{l^2 - m^2}, \qquad i\Delta_\psi(l) = \frac{i}{l^2 - M^2}. \qquad (4.5)$$

The solution of the BSE is given by

$$T(s) = \frac{g}{1 - gG(s)} = g + gG(s)g + gG(s)gG(s)g + \ldots \qquad (4.6)$$

which is interpreted as a summation of an infinite series of s-channel loop graphs, the so-called bubble chain (or bubble sum). From inversion of eq. (4.6) it immediately follows that

$$\operatorname{Im} T^{-1} = -\operatorname{Im} G = \frac{|\mathbf{q}_{\rm cm}|}{8\pi\sqrt{s}}\,\theta(s - (m+M)^2), \qquad (4.7)$$

which amounts to the statement of two-particle unitarity for the scattering amplitude. Strictly speaking, the integral G is divergent, but its divergent piece can be absorbed in the coupling g, as demonstrated in eq. (1.48). In the following, we will thus assume that the scalar loop integral G has been rendered finite by an appropriate redefinition of the coupling and omit the divergent pieces henceforth.

Let us now turn to the coupling of an external photon field to the solution of the BSE. By minimal substitution

$$\begin{aligned}\partial_\mu &\to \nabla_\mu = \partial_\mu + ie_\phi \mathcal{A}_\mu \qquad \text{for } \phi, \\ \partial_\mu &\to \nabla_\mu = \partial_\mu + ie_\psi \mathcal{A}_\mu \qquad \text{for } \psi,\end{aligned} \qquad (4.8)$$

where $\mathcal{A}_\mu$ is the photon field and e_ϕ (e_ψ) the charge of the scalar field ϕ (ψ), one obtains a locally gauge invariant Lagrangian. The coupling of the photon with incoming momentum k to the four external legs of a bubble chain leads to the amplitudes

$$\begin{aligned}T_1^\mu &= e_\phi\, T(s')\,\Delta(p_1+k)\,(2p_1+k)^\mu, \\ T_2^\mu &= e_\psi\, T(s')\,\Delta(p_2+k)\,(2p_2+k)^\mu, \\ T_3^\mu &= e_\phi(2p_3-k)^\mu\,\Delta(p_3-k)\,T(s), \\ T_4^\mu &= e_\psi(2p_4-k)^\mu\,\Delta(p_4-k)\,T(s),\end{aligned} \qquad (4.9)$$

where $s' = (p+k)^2$. Multiplying these contributions with the four-momentum of the photon, k_μ, and setting the external legs on-shell yields

$$(e_\phi + e_\psi)\,[T(s') - T(s)], \qquad (4.10)$$

which in general does not vanish. This underlines that, in order to achieve gauge invariance, it is not sufficient to couple the photon only to external legs. One rather has to include contributions which arise due to the coupling of the photon to intermediate states within the bubble chain, leading to the additional contributions

$$\begin{aligned}T_5^\mu &= ie_\phi T(s') \int_l \Delta_\phi(l+k)\,(2l+k)^\mu\,\Delta_\phi(l)\,\Delta_\psi(p-l)\,T(s), \\ T_6^\mu &= ie_\psi T(s') \int_l \Delta_\psi(l+k)\,(2l+k)^\mu\,\Delta_\psi(l)\,\Delta_\phi(p-l)\,T(s).\end{aligned} \qquad (4.11)$$

By employing the following algebraic identity, equivalent to the Ward-Takahashi identity at lowest order in e and g,

$$k_\mu (2l+k)^\mu = \Delta^{-1}_{\phi/\psi}(l+k) - \Delta^{-1}_{\phi/\psi}(l) \qquad (4.12)$$

it is straightforward to show that

$$k_\mu \left(T_5^\mu + T_6^\mu\right) = (e_\phi + e_\psi)\, T(s')\left[G(s) - G(s')\right] T(s)\,. \qquad (4.13)$$

The last expression can be rewritten by making use of the BSE

$$T(s')[G(s)-G(s')]T(s) = T(s')\left[g^{-1}T(s) - 1\right] - \left[T(s')g^{-1} - 1\right]T(s) = T(s) - T(s')\,. \qquad (4.14)$$

Adding up all contributions we arrive at

$$k_\mu \sum_{i=1}^{6} T_i^\mu = 0\,, \qquad (4.15)$$

which confirms the gauge invariance of the photon coupling to the bubble chain.

At the same time, insertion of the photon coupling at all possible places in the bubble chain guarantees unitarity of the scattering matrix up to radiative corrections of order $\mathcal{O}(e^3)$. To this end, we remark that in the transition $\phi\psi\gamma \to \phi\psi$ we can restrict ourselves to the states $|\phi\psi\rangle$ and $|\phi\psi\gamma\rangle$. Unitarity of the $\mathcal{S}$-matrix, $\mathcal{S} = 1 - i\mathcal{T}$, implies then (compare eq. (1.39))

$$\langle\phi\psi|\mathcal{T}-\mathcal{T}^\dagger|\phi\psi\gamma\rangle = -i \int_{PS} \left\{ \langle\phi\psi|\mathcal{T}^\dagger|\phi'\psi'\rangle\langle\phi'\psi'|\mathcal{T}|\phi\psi\gamma\rangle + \langle\phi\psi|\mathcal{T}^\dagger|\phi'\psi'\gamma'\rangle\langle\phi'\psi'\gamma'|\mathcal{T}|\phi\psi\gamma\rangle \right\}, \qquad (4.16)$$

where $\int_{PS}$ denotes the phase space integral for the set of intermediate states $|\phi'\psi'\rangle$ and $|\phi'\psi'\gamma'\rangle$. We have introduced a superscript for the intermediate particles with running three-momenta, ϕ', ψ' and γ', in order to distinguish them from the external particles ϕ, ψ and γ with fixed momenta. Note that the last matrix element $\langle\phi'\psi'\gamma'|\mathcal{T}|\phi\psi\gamma\rangle$ contains a disconnected piece of order $\mathcal{O}(e^0)$ in which the photon does not couple to the bubble chain and thus appears in the $\mathcal{O}(e)$ part of the unitarity relation. The remaining connected diagrams of this matrix element which contain the coupling of the photon to the bubble chain are of order $\mathcal{O}(e^2)$ and will be neglected in the following. By making use of the symmetry $\langle i|\mathcal{T}|j\rangle = \langle j|\mathcal{T}|i\rangle$ due to time reversal invariance, eq. (4.16) can be rewritten as

$$-2\,\mathrm{Im}\langle\phi\psi|\mathcal{T}|\phi\psi\gamma\rangle = \int_{PS} \left\{ \langle\phi\psi|\mathcal{T}|\phi'\psi'\rangle^* \langle\phi'\psi'|\mathcal{T}|\phi\psi\gamma\rangle + \langle\phi\psi|\mathcal{T}|\phi'\psi'\gamma\rangle^* \langle\phi'\psi'\gamma|\mathcal{T}|\phi\psi\gamma\rangle \right\}, \qquad (4.17)$$

where now the phase space integral applies only to the particles ϕ' and ψ'. The last equation represents the Cutkosky cutting rules [103]. At the diagrammatic level this amounts to cutting a pair of ϕ and ψ propagators at all possible places in the bubble chain (keeping in mind that the photon is merely an external particle). The two terms on the right hand side represent the two possibilities for the photon to couple to the bubble chain before or after the cut. Since these are the only possible cuts in the bubble chain leading to imaginary pieces in the relevant kinematic region, one verifies eq. (4.17) and hence unitarity of the $\mathcal{S}$-matrix up to radiative corrections. However, if the photon couples to a propagator, there is also the possibility to cut in the corresponding diagram both propagators which are directly connected to the photon.

For a photon with $k^2 < 4\min(m^2, M^2)$ which is the case both for physical photons and for virtual photons from electron scattering, these cuts do not yield imaginary values and can be safely omitted here.

Having convinced ourselves that it is possible to obtain gauge invariant and (up to radiative corrections) unitary amplitudes by taking into account the coupling of the photon to scalar fields in all possible ways in the bubble chain, we can now continue by applying this procedure to the slightly more complicated case of the chiral effective meson-baryon Lagrangian.

4.3 Weinberg-Tomozawa term

The chiral effective Lagrangian describing the interactions between the octet of Goldstone bosons (π, K, η) and the ground state baryon octet $(N, \Lambda, \Sigma, \Xi)$ is given at leading chiral order by eq. (1.33),

$$\mathcal{L}_{MB}^{(1)} = i\langle \bar{B}\gamma_\mu [D^\mu, B]\rangle - m_0 \langle \bar{B}B \rangle + \ldots . \tag{4.18}$$

We have only displayed the terms relevant for the present investigation and omitted the two operators which contain the axial vector couplings of the mesons to the baryons. In the present investigation, we restrict ourselves to interaction kernels of the BSE given by contact interactions. The axial vector couplings of the mesons to the baryons could in principle contribute via direct and crossed Born terms, but the crossed Born term corresponds to three-body intermediate states which are beyond the scope of this work, and we neglect the Born terms throughout. (The inclusion of Born terms in the interaction kernel is deferred to future work.) In fact, many coupled-channels approaches only take into account the contact interaction originating from the Lagrangian in (4.18), see e.g. [104] and references therein.

The covariant derivative of the baryon field was defined in eq. (1.34). Here we only include the external vector field in the chiral connection, so that

$$\Gamma_\mu = \frac{1}{2}[u^\dagger, \partial_\mu u] - \frac{i}{2}\left(u^\dagger v_\mu u + u v_\mu u^\dagger\right). \tag{4.19}$$

More specifically, we set $v_\mu = -e\mathcal{A}_\mu Q$, where $Q = \frac{1}{3}\text{diag}(2, -1, -1)$ is the quark charge matrix. Expansion of the chiral connection in the meson fields ϕ yields at leading order a $\phi^2 \bar{B}B$ contact interaction, the Weinberg-Tomozawa term, which we choose to be the driving term for the BSE in this section.

We also recall from eq. (1.18) the mesonic piece of the Lagrangian at leading chiral order,

$$\mathcal{L}_\phi^{(2)} = \frac{F^2}{4}\langle \nabla_\mu U^\dagger \nabla^\mu U \rangle + \frac{F^2}{4}\langle \chi_+ \rangle, \tag{4.20}$$

where we used the definition $\chi_+ = 2B(u^\dagger \mathcal{M} u^\dagger + u\mathcal{M}u)$ as given in eq. (3.5). Neglecting the axial-vector field a, the covariant derivative of the meson fields is given by

$$\nabla_\mu U = \partial_\mu U - iv_\mu U + iU v_\mu . \tag{4.21}$$

In the Bethe-Salpeter formalism we choose to work with the propagators

$$\Delta_i(p) = \frac{1}{p^2 - M_i^2},$$

$$S_a(p) = \frac{1}{\not{p} - m_a}, \tag{4.22}$$

with flavor indices i, a and physical meson and baryon masses, M_i and m_a, respectively. However, the following calculations are valid for all propagators satisfying the Ward-Takahashi identities

$$k^\mu V_\mu^\phi(p+k, k) = \Delta^{-1}(p+k) - \Delta^{-1}(p),$$
$$k^\mu V_\mu^B(p+k, k) = S^{-1}(p+k) - S^{-1}(p), \quad (4.23)$$

where V_μ^ϕ (V_μ^B) are the corresponding $\gamma\phi^2$ ($\gamma\bar{B}B$) three-point functions, with the charge prefactors omitted (compare eq. (4.12)).

In the presence of a general interaction kernel A and coupled channels consisting of a meson-baryon pair the BSE for the process $\phi(q_i)B(p_i) \to \phi(q_f)B(p_f)$ generalizes to

$$T_{fi}(p; q_f, q_i) = A_{fi}(p; q_f, q_i) + i \sum_l \int_k T_{fl}(p; q_f, k) \Delta_j(k) S_a(p+k) A_{li}(p; k, q_i)$$
$$= A_{fi}(p; q_f, q_i) + i \sum_l \int_k A_{fl}(p; q_f, k) \Delta_j(k) S_a(p+k) T_{li}(p; k, q_i) \quad (4.24)$$

with $p = p_i + q_i = p_f + q_f$. The index $l = \{\phi_j, B_a\}$ runs over the channels which couple both to the initial and final state, i and f. In the second line, we have made use of the symmetry property of the BSE, which is easily verified by solving the integral equation iteratively, to an arbitrary order in the kernel A. Also note that we have replaced the common mass of the ground state baryon octet, m_0, by the physical baryon masses m_a. This is consistent with the chiral order of the interaction kernel derived from the Weinberg-Tomozawa term and, in particular, produces the unitarity cuts at the physical thresholds.

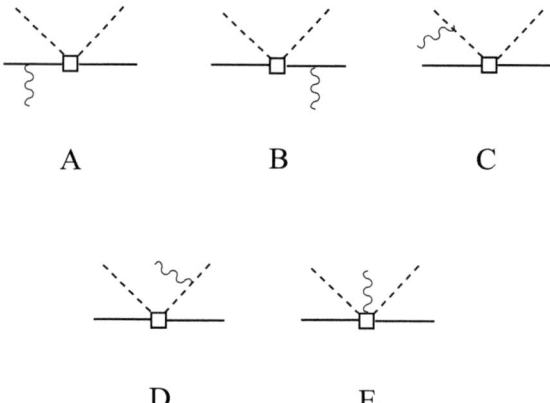

Figure 4.1: Tree diagrams for the process $\gamma\phi B \to \phi B$. Solid, dashed and wavy lines correspond to baryons, mesons and photons, respectively. The square denotes the vertex from the leading order Lagrangian.

After setting up the formalism we first calculate the tree level contributions for the coupling of a photon to meson-baryon scattering. The pertinent Feynman diagrams for the process

$\gamma(k)\phi_i(q_i)B_a(p_i) \to \phi_j(q_f)B_b(p_f)$ are depicted in Figure 4.1. In addition to the coupling of the photon to the propagators the chiral connection in eq. (4.18) gives rise to a $\gamma\phi^2\bar{B}B$ vertex, fig. 4.1 E.

The tree contributions to the transition amplitude read

$$\begin{aligned}T_\mu^{(\text{tree})\,bj,ai} = -\frac{e}{4F^2}\Big\{&(\not{q}_i+\not{q}_f)S_a(p_i+k)\gamma_\mu \hat{Q}^a + \gamma_\mu \hat{Q}^b S_b(p_f-k)(\not{q}_i+\not{q}_f)\\&+(\not{q}_i+\not{q}_f+\not{k})\Delta_i(q_i+k)[2q_i+k]_\mu \hat{Q}^i\\&+(\not{q}_i+\not{q}_f-\not{k})\Delta_j(q_f-k)[2q_f-k]_\mu \hat{Q}^j\\&-\gamma_\mu(\hat{Q}^j+\hat{Q}^i)\Big\}\langle\lambda^{b\dagger}[[\lambda^{j\dagger},\lambda^i],\lambda^a]\rangle\,,\end{aligned}\quad(4.25)$$

where $\hat{Q}^a\lambda^a = [Q,\lambda^a]$ (no summation over a) is the charge of the particle a in units of e and the λ^i are the generators of the $SU(3)$ Lie-Algebra in the physical basis. By multiplying the tree contributions in eq. (4.25) with k_μ gauge invariance is easily verified, if the momenta of the particles are put on-shell.

In order to prove gauge invariance for the coupling of the photon to the bubble chain, it is convenient to consider first the diagrams presented in fig. 4.2 with the pertinent contributions given by (a,b,c,d denote baryon flavor indices, whereas i,j,m,n represent meson flavors):

Fig. 4.2A:
$$A_\mu^{bj,ai} = \frac{e}{4F^2}\gamma_\mu(\hat{Q}^j+\hat{Q}^i)\langle\lambda^{b\dagger}[[\lambda^{j\dagger},\lambda^i],\lambda^a]\rangle\,,\quad(4.26)$$

Fig. 4.2B:
$$i^2\int_l\int_q T^{bj,dn}(p';q_f,q)\,S_d(p'-q)\Delta_n(q)\,A_\mu^{dn,cm}S_c(p-l)\Delta_m(l)\,T^{cm,ai}(p;l,q_i)\,,\quad(4.27)$$

Fig. 4.2C:
$$i\int_l T^{bj,cm}(p';q_f,l)\,S_c(p'-l)\Delta_m(l)\,A_\mu^{cm,ai}\,,\quad(4.28)$$

Fig. 4.2D:
$$i\int_l A_\mu^{bj,cm}S_c(p-l)\Delta_m(l)\,T^{cm,ai}(p;l,q_i)\,,\quad(4.29)$$

Fig. 4.2E:
$$i^2\int_l T^{bj,cm}(p';q_f,l+k)\,S_c(p-l)\Delta_m(l+k)\,(-ie\hat{Q}^m(2l+k)_\mu)\Delta_m(l)\,T^{cm,ai}(p;l,q_i)\,,\quad(4.30)$$

Fig. 4.2F:
$$i^2\int_l T^{bj,cm}(p';q_f,l)\,S_c(p'-l)\,(-ie\hat{Q}^c\gamma_\mu)S_c(p-l)\Delta_m(l)\,T^{cm,ai}(p;l,q_i)\,,\quad(4.31)$$

with $p'=p+k=p_i+q_i+k=p_f+q_f$.

For general meson momenta p,q (i.e. not necessarily on-shell), the quantity $A_\mu^{bj,ai}$ satisfies the relation

$$k^\mu A_\mu^{bj,ai} = e\Big\{(\hat{Q}^b-\hat{Q}^a)A^{bj,ai}(q,p) - \hat{Q}^i A^{bj,ai}(q,p+k) + \hat{Q}^j A^{bj,ai}(q-k,p)\Big\}\,,\quad(4.32)$$

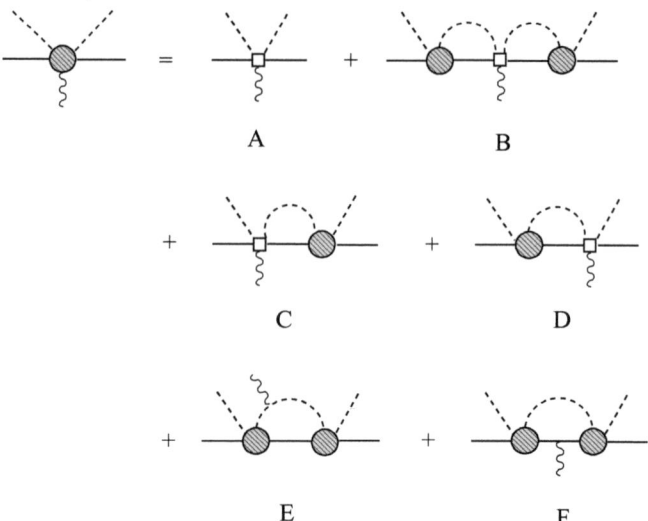

Figure 4.2: Bubble chain diagrams for the process $\gamma\phi B \to \phi B$. Solid, dashed and wavy lines correspond to baryons, mesons and photons, respectively. The square denotes the vertex from the leading order Lagrangian, the filled circle represents the bubble chain derived from the BSE.

where A is the amplitude deduced from the Weinberg-Tomozawa term

$$A^{bj,ai}(q,p) = -\frac{1}{4F^2}(\slashed{q}+\slashed{p})\langle\lambda^{b\dagger}[[\lambda^{j\dagger},\lambda^i],\lambda^a]\rangle\,. \qquad (4.33)$$

Making extensively use of eq. (4.32) and the BSE it is straightforward to show that the contributions of eqs. (4.26)-(4.31) multiplied by k_μ yield in total

$$e\left\{-\hat{Q}^a T^{bj,ai}(p';q_f,q_i)+\hat{Q}^b T^{bj,ai}(p;q_f,q_i)-\hat{Q}^i T^{bj,ai}(p';q_f,q_i+k)+\hat{Q}^j T^{bj,ai}(p;q_f-k,q_i)\right\}\,. \qquad (4.34)$$

This compensates exactly the contributions from the remaining four diagrams where the photon couples to the external on-shell legs of the bubble chain. We have thus confirmed that gauge invariance is achieved if all possible diagrams of a photon coupling to a bubble chain are taken into account. In particular, it is not sufficient to consider only the coupling of the photon to external legs, since this will lead to a gauge dependent amplitude. On the other hand, within the field theoretical framework applied here it is also not sufficient to take into account only the coupling of the photon to the interaction kernel. Note that in the proof given here we do not assume the so-called on-shell approximation for the interaction kernel.

It is also important to stress that an explicit evaluation of the BSE was not necessary and thus we do not need to specify the regularization scheme to render the loop integral in the BSE finite. Any regularization procedure which satisfies the Ward identities both for the propagators, eq. (4.23), and the interaction kernel, eq. (4.32), will maintain gauge invariance in the BSE as outlined in the proof.

Finally, the unitarity constraint for the full amplitude T_μ can be proven in analogy to the treatment in sec. 4.2. The detailed derivation of unitarity is deferred to sec. 4.5. Thus, also in order to obtain a unitarized amplitude (up to radiative corrections) one must take into account the coupling of the photon at all possible places in the bubble chain.

4.4 Higher order interaction kernels

In this section, we will consider more complicated structures for the interaction kernel as they arise at higher chiral orders in the effective Lagrangian. Since we restrict ourselves to $\phi^2 \bar{B} B$ contact interactions, the most general form of a term without the chiral invariant field strength tensor $F^+_{\mu\nu}$ (see eq. (1.38)) is given by

$$\mathcal{L}_{\text{int}} = \bar{B}\, C_{\mu_1 \cdots \mu_l \mu_{l+1} \cdots \mu_m \mu_{m+1} \cdots \mu_n} \left(D^{\mu_1} \ldots D^{\mu_l} \tilde{\phi} \right) \left(D^{\mu_{l+1}} \ldots D^{\mu_m} \phi^\dagger \right) \left(D^{\mu_{m+1}} \ldots D^{\mu_n} \tilde{B} \right), \quad (4.35)$$

where we have suppressed flavor indices for brevity (we merely kept the symbol "$\sim$" as a reminder of the flavor structure indicating that the in- and outgoing baryons and mesons can be different) and introduced the notation $D_\mu = \partial_\mu + i\hat{e}\mathcal{A}_\mu$ with $\hat{e}$ the charge of the particle D_μ is acting on. As the contact interactions originate from a gauge invariant Lagrangian, charge conservation is guaranteed at each vertex. The introduction of explicit flavor indices does not change any of the following conclusions. Note that the constant C in eq. (4.35) may also contain elements of the Clifford algebra. From this Lagrangian one derives both a $\tilde{\phi}(\tilde{q})\tilde{B}(\tilde{p}) \rightarrow \phi(q)B(p)$ contact interaction $A(\tilde{p}, \tilde{q}, q)$ and a $\gamma(k)\tilde{\phi}(\tilde{q})\tilde{B}(\tilde{p}) \rightarrow \phi(q)B(p)$ vertex $\epsilon_\mu A^\mu(\tilde{p}, \tilde{q}, q, k)$, where ϵ_μ is the polarization vector of the photon. Due to the form of the contact term (4.35), which can always be obtained by partial integration, the vertices do not depend explicitly on the momentum p. In app. D it is shown that they satisfy the relation

$$k_\mu A^\mu(p_i, q_i, q_f, k) = \hat{e}_\phi A(p_i, q_i, q_f - k) - \hat{e}_{\tilde{\phi}} A(p_i, q_i + k, q_f)$$
$$+ \hat{e}_B A(p_i, q_i, q_f) - \hat{e}_{\tilde{B}} A(p_i + k, q_i, q_f) \quad (4.36)$$

for general momenta of the particles. This equation is the analog of eq. (4.32) for the Weinberg-Tomozawa term. It follows then immediately by applying the same arguments as in the previous section that the coupling of the photon to the two-particle state of the BSE yields a gauge invariant amplitude also in the presence of more complicated contact interactions of the type eq. (4.35).

Note that throughout we have not considered the dimension two magnetic moment coupling $\sim \sigma^{\mu\nu} F^+_{\mu\nu}$ and higher order operators involving the chiral covariant field strength tensor $F^+_{\mu\nu}$. Such terms are of course present in the effective Lagrangian and must be considered at the appropriate order in the chiral expansion of the interaction kernel. However, these are of the form $\partial_\nu v_\mu (\mathcal{O}^{\nu\mu} - \mathcal{O}^{\mu\nu})$ with some operator $\mathcal{O}^{\nu\mu}$ and the pertinent vertex in momentum space vanishes upon contraction with the photon momentum k_μ.

As already mentioned in the previous section, the proof of gauge invariance does not depend on the specific choice of the meson and baryon propagators, but is rather valid for all propagators satisfying the Ward-Takahashi identities with the corresponding $\gamma \phi^2$ and $\gamma \bar{B} B$ three-point functions. For example, one can define the BSE by employing propagators with the physical masses for the intermediate states and derive the interaction kernel from the effective Lagrangian to a given chiral order. This automatically produces the correct physical thresholds of the unitarity cuts and we have followed this path in the present investigation.

Alternatively, one may prefer to deduce the propagators from the effective Lagrangian as well. To leading chiral order this implies a common baryon octet mass shifting the threshold of the unitarity cuts to unphysical values. At higher chiral orders the inclusion of self-energy diagrams for the meson and baryon propagators will cure the situation by restoring the physical thresholds. The self-energy diagrams will modify the simple form of the propagators given in eq. (4.24); e.g., the meson propagators will acquire the form

$$\Delta(p) = \frac{Z}{p^2 - M^2 - \Sigma_R(p)} \quad (4.37)$$

with M being the physical meson mass, Z the appropriate wave function renormalization constant, and $\Sigma_R(p)$ the renormalized self-energy.

In order to prove gauge invariance in the latter approach, one must also take into account the coupling of the photon to the self-energy corrections of the propagators. Since the effective Lagrangian is gauge invariant, the corresponding propagators and three-point functions satisfy the pertinent Ward-Takahashi identities and the proof is equivalent to the one given in the previous section.

We also would like to compare the present investigation with the work of [94, 95]. Although similar in spirit, the authors study therein an integral equation for the two-body Green's function, whereas we prefer to work with an equation for the scattering amplitude. "Gauging" the integral equation for the Green's function as outlined in [94, 95], i.e. adding a vector index μ to all the terms of the equation such that a linear equation in μ-labeled quantities results, amounts to attaching an external photon everywhere including the external legs. In our framework, the straightforward application of the gauging method to the integral equation for the scattering amplitude fails, as in this case the procedure does not yield the contributions where the photon couples to the external legs. (Of course, one could correct this by adding the missing contributions by hand.) On the other hand, our approach is more convenient within the chiral effective framework, as it allows a direct comparison with the scattering amplitude derived in the perturbative scheme of ChPT. Moreover, the authors of [94, 95] restrict themselves merely to one two-body channel, whereas in the present investigation this is generalized to several coupled channels. In contrast to [94, 95] we explicitly specify the interaction kernel by deriving the vertices from the chiral effective Lagrangian and utilizing them as interaction kernels in the BSE.

4.5 Unitarity

After having constructed a gauge invariant amplitude for $B\phi\gamma \to B\phi$ with Weinberg-Tomozawa or more general contact interaction kernels, we would like to investigate unitarity of the obtained amplitude. The calculation presented in this section generalizes the findings for the scalar field theory presented at the end of sec. 4.2, as one must take care of the non-commutative nature of the matrix amplitudes due to the Clifford algebra and coupled channels. Moreover, we do not assume symmetry of the transition amplitudes under exchange of initial and final states. In operator form the statement of a unitary scattering matrix amounts to

$$\mathcal{T} - \mathcal{T}^\dagger = -i\mathcal{T}^\dagger\mathcal{T} \; . \quad (4.38)$$

For brevity we introduce a short-hand notation for the BSE, eq. (4.24),

$$T(p) = A + \int T(p)G(p)A = A + \int AG(p)T(p), \tag{4.39}$$

where p is the external momentum and $G = iS\Delta$. The BSE for the meson-baryon scattering amplitude T is easily transformed into the unitarity relation

$$T - \bar{T} = \int \bar{T}(G - \bar{G})T \tag{4.40}$$

with $\bar{O} \equiv \gamma_0 O^\dagger \gamma_0$, as both T and G are elements of the Clifford algebra. Note that the adjoint $O^\dagger$ also implies taking the transposed matrix in channel space. We will see that the quantity $G - \bar{G}$ is equal to setting the intermediate meson-baryon pairs on-shell in eq. (4.40). For invariant energies below the lowest three-particle threshold eq. (4.40) is thus equivalent to the unitarity constraint (4.38). The reader should compare this also with eq. (1.41) for the case of spinless particles.

If T is an analytic function, one can apply the residue theorem and rewrite the difference $G - \bar{G}$ as

$$i(S(p-l)\Delta(l) + \bar{S}(p-l)\bar{\Delta}(l)) \rightarrow i(-2\pi i)^2 \delta_+(l^2 - M^2)\delta_+((p-l)^2 - m^2)\left[\not{p} - \not{l} + m\right], \tag{4.41}$$

where l and p are the loop and external momentum, respectively, and $\delta_+(k^2 - \mu^2) = \delta(k^2 - \mu^2)\theta(k_0)$. For a detailed derivation of this replacement, see e.g. sec. 7.3 of [18]. The last equation can even be generalized to non-analytic T which does not have coinciding poles with G. In particular, the above replacement is valid if T describes the solution of the BSE with polynomial interaction kernels, as can be seen by insertion of its defining equation (4.24) into eq. (4.40), making use of $\bar{A} = A$. Schematically, the calculation is performed as follows:

$$\int \bar{T}(G - \bar{G})T = \int \bar{T}GT - \int \bar{T}\bar{G}\left(A + \int AGT\right)$$

$$= \int \bar{T}GT - (\bar{T} - A) - \int\int \bar{T}\bar{G}AGT$$

$$= \int \bar{T}GT + A - \bar{T} - \int (\bar{T} - A)GT = T - \bar{T}.$$

In order to prove unitarity for the transition $B\phi\gamma \rightarrow B\phi$, it is convenient to introduce the amplitude

$$M^\mu_{\phi\gamma} = V^\mu_{disc} + T^\mu \tag{4.42}$$

with the disconnected piece

$$V^\mu_{disc} = 2E_{q_i}(2\pi)^3\delta^{(3)}(\boldsymbol{q}_i - \boldsymbol{q}_f)V^\mu_B + 2E_{p_i}(2\pi)^3\delta^{(3)}(\boldsymbol{p}_i - \boldsymbol{p}_f)V^\mu_\phi \tag{4.43}$$

and T^μ the transition amplitude calculated in sections 4.3 and 4.4. The energies of the particles are given by $E_{q_i} = \sqrt{q_i^2 + M_\phi^2}$ and $E_{p_i} = \sqrt{p_i^2 + m_B^2}$, respectively. The piece V^μ_{disc} is represented by the two disconnected diagrams in which the photon couples either to the baryon or meson, while the other particle does not interact at all. Although V^μ_{disc} does not contribute to on-shell matrix elements and could in principle be omitted in the unitarity relation, its introduction generalizes unitarity beyond the physical region. From the amplitude $M^\mu_{\phi\gamma}$ one constructs the reversed amplitude $M^\mu_{\gamma\phi}$ for the process $B(p_f)\phi(q_f) \to B(p_i)\phi(q_i)\gamma(k)$. Neglecting radiative corrections and below the lowest three-particle (i.e. baryon two-meson) threshold unitarity implies

$$M^\mu_{\phi\gamma} - \bar{M}^\mu_{\gamma\phi} = \int \bar{T}(p')(G(p') - \bar{G}(p'))M^\mu_{\phi\gamma} + \int \bar{M}^\mu_{\gamma\phi}(G(p) - \bar{G}(p))T(p) , \qquad (4.44)$$

where we have replaced again the two-body phase space integration by the four dimensional integral over $G - \bar{G}$. Note that in the physical region $M^\mu_{\phi\gamma}$ reduces to T^μ.

By inserting the amplitudes T^μ and T from the BSE and making use of the unitarity statement for T, eq. (4.40), as well as $\bar{V}^\mu_{\phi/B} = V^\mu_{\phi/B}$ and $\bar{A}^\mu = A^\mu$ from eq. (4.26), one can indeed confirm the unitarity constraint for $M^\mu_{\phi\gamma}$ and thus for T^μ for on-shell matrix elements.

We refrain from presenting the entire and tedious calculation here, but would like to comment on two points. First, the contribution of the disconnected graphs drops out on the l.h.s. of the unitarity statement (4.44) (due to the symmetry of these graphs under interchange of incoming and outgoing particles and $\bar{V}^\mu_{\phi/B} = V^\mu_{\phi/B}$). On the r.h.s., they produce terms of the type

$$\int_l \bar{T}(p') \, (G(p') - \bar{G}(p')) \, 2E_{q_i}(2\pi)^3\delta^{(3)}(\mathbf{q}_i - \mathbf{l})V^\mu_B$$
$$= \int_l \bar{T}(p') \, i(-2\pi i)^2 \, \delta_+((p'-l)^2 - m^2) \, \delta_+(l^2 - M^2) \, [\slashed{p}' - \slashed{l} + m] \, 2E_{q_i}(2\pi)^3\delta^{(3)}(\mathbf{q}_i - \mathbf{l})V^\mu_B$$
$$= \int d^3l \, \bar{T}(p')(-2\pi i) \, \delta_+((p'-l)^2 - m^2) \, [\slashed{p}' - \slashed{l} + m] \, \delta^{(3)}(\mathbf{q}_i - \mathbf{l})V^\mu_B$$
$$= \bar{T}(p')(-2\pi i) \, \delta_+((p' - q_i)^2 - m^2)[\slashed{p}' - \slashed{q}_i + m] \, V^\mu_B$$
$$= \bar{T}(p') \, [S(p_i + k) - \bar{S}(p_i + k)] \, V^\mu_B. \qquad (4.45)$$

For on-shell matrix elements (and $k \neq 0$) this vanishes as $\epsilon \to 0$ in the propagators, but it happens that all terms of this type cancel each other on the r.h.s. of eq. (4.44) even for general external momenta. Our second comment concerns some contributions from diagrams 4.2E and F. On the l.h.s. of eq. (4.44) one obtains, e.g., the combination

$$i\int_l \bar{T}(p') \, S(p'-l) \, \Delta(l) \, V^\mu_B \, S(p-l) \, T(p) + i\int_l \bar{T}(p') \, \bar{S}(p'-l) \, \bar{\Delta}(l) \, V^\mu_B \, \bar{S}(p-l) \, T(p) \qquad (4.46)$$

which represents the discontinuity of the transition amplitude stemming from fig. 4.2F (the two integrals only differ in the sign of the "$i\epsilon$" terms in the propagators). According to the Cutkosky cutting rules and for momenta $k^2 < 4m^2$ this discontinuity is given by

$$\int_l \bar{T}(p') \, (G(p') - \bar{G}(p')) \, V^\mu_B \, S(p-l) \, T(p) + \int_l \bar{T}(p') \, \bar{S}(p'-l) \, V^\mu_B \, (G(p) - \bar{G}(p)) \, T(p). \qquad (4.47)$$

In the kinematical region which is of relevance here the cut through the two propagators connected to the photon does not contribute to the discontinuity and can be safely neglected. We conclude by emphasizing that the unitarity constraint (4.44) is only fulfilled if the photon couples to all possible places in the bubble chain.

4.6 Conclusions

In the present chapter, we have studied how gauge invariance is obtained for a photon coupling to a two-body state described by the solution of the Bethe-Salpeter equation. We have discussed the procedure both for a simple complex scalar field theory and for interaction kernels derived from chiral effective Lagrangians in the meson-baryon sector. In the latter case, we have first considered the Weinberg-Tomozawa term and afterwards the most general contact interaction consisting of two mesons and two baryons which can arise in the chiral effective framework. Our study underlines that it is not sufficient to take into account only the coupling of the photon to the external legs as has been done in many calculations based on chiral unitary approaches, but one rather has to include all possible contributions of the photon coupling to the vertices and intermediate states. Neither is it sufficient to consider only the coupling of the photon to the interaction kernel. At the same time, coupling of the photon to the bubble chain at all possible places is necessary, in order to guarantee a unitary scattering matrix up to radiative corrections.

For the interaction kernels discussed in the present work we have shown explicitly that gauge invariance is maintained in this manner. It is accomplished without assuming the on-shell approximation for the interaction kernel. Moreover, the explicit evaluation of the loop integral in the Bethe-Salpeter equation is not necessary and hence the proof does not depend on the chosen regularization scheme. But the regularization procedure is required to be consistent with the Ward identities both for the propagators and the interaction kernels. This study will also be of importance for photo- and electroproduction processes of mesons on nucleons and for radiative decays of baryons and mesons which must be treated in a similar way, in order to achieve gauge invariant and unitarized amplitudes. One example for an application will be presented in the next chapter. The method outlined in this chapter has also been applied to the study of the electromagnetic mean squared radii of the $\Lambda(1405)$ in [105].

Chapter 5

Threshold kaon photo-and electroproduction

5.1 Introduction

In the foregoing chapter, we have developed a method that allows to implement the electromagnetic interaction in a given unitarized amplitude, in a way that is in accord with both gauge invariance and unitarity. More specifically, we started with amplitudes for elastic two-particle scattering, which were constructed by means of a Bethe-Salpeter equation (BSE) and therefore obeyed the requirement of two-body unitarity, see eqs. (4.7) and (4.40). Then we coupled a photon to the sum of bubble graphs generated by the BSE, employing the prescription (familiar from quantum field theory textbooks) that the photon has to be attached at every position in the graphs, wherever this is possible. We also showed that the result of this procedure is an amplitude for the absorption of the photon by the original two-particle system which again obeys the strictures of unitarity (projecting on the subspace of two-particle states as usual).

It is the purpose of this chapter to apply this method to a specific process, namely, kaon photo- and electroproduction off protons. Let us make some general remarks to motivate the selection of this specific process.

The hadronic spectrum is still the least understood property of QCD. Most theoretical models predict much more states than are actually observed so far in experiments, see e.g. [106–108]. This is sometimes called the 'missing resonance problem'. The search for missing resonances has therefore been an important goal of various experimental efforts. The properties of baryon resonances are presently under thorough investigation at several facilities, e.g., at ELSA, JLab, MAMI, GRAAL, COSY and SPring-8. Due to their hadronic decay modes, however, many baryon resonances have large overlapping widths which makes it a difficult task to study individual states. In this respect, polarization observables can be used as a tool to filter out specific resonances in specific reactions.

A possible explanation of the missing resonances could be that these states do not couple strongly to the pion-nucleon channels which have provided to a large extent the resonance data. A strong coupling of these resonances to channels with strange particles could be unraveled in photoproduction processes of $K\Lambda$ and $K\Sigma$. Such experiments have been recently undertaken at SAPHIR [109,110] and CLAS [111,112] with high precision. More recently, $K^0\Sigma^+$ has been measured with the CB/TAPS detector at ELSA [113]. These data indicate that resonances so far undetected might have been observed, but an unambiguous theoretical interpretation is still lacking. In addition, the beam polarization asymmetry for $\gamma p \to K^+\Lambda, K^+\Sigma^0$ has been investigated by the LEPS collaboration at SPring-8 [114].

Based on an analysis of cross sections, beam asymmetries, and recoil polarizations, the Bonn-Gatchina resonance model, e.g., demands among others the $P_{11}(1840)$, $D_{13}(1870)$, $D_{13}(2170)$ resonances [115, 116], whereas in the model of [117] a $P_{13}(1830)$ is preferred.

Electroproduction experiments are even more sensitive to the structure of the nucleon due to the longitudinal coupling of the virtual photon to the nucleon spin and might in addition yield some insight into the possible onset of perturbative QCD, see e.g. [118, 119]. But at low photon virtuality experimental data for $\gamma p \to K\Lambda, K\Sigma$ are still not available.

Apparently, there are sufficient data to be met by theoretical approaches. In addition to the photoproduction data, further experimental constraints are provided by pion-induced reactions on the proton. Any theoretical approach that aims to describe the photoproduction data ought to be consistent with the corresponding pion-induced data.

A successful theoretical approach to meson-baryon scattering is provided by chiral unitary methods, see e.g. [56, 57, 104, 120, 121]. In this framework the chiral effective Lagrangian is utilized to derive, for example, the interaction kernel in a Bethe-Salpeter equation (BSE) which iterates meson-baryon rescattering to infinite order. The BSE generates resonances dynamically, hence, without their explicit inclusion the importance of resonances can be studied. Chiral unitary approaches have been implemented quite successfully for photoproduction processes, see e.g. refs. [89–91,122–124], but as a simplification only those diagrams were taken into account where the photon is absorbed first and then the produced meson-baryon pair undergoes final state interaction. This simplified treatment violates, in general, gauge invariance. As we have seen in the previous chapter, diagrams with the photon coupling to any intermediate state of the meson-baryon bubble chain must be taken into account in order to guarantee gauge invariance. One goal of the analysis in the present chapter is to study the importance of these additional contributions which render the amplitude gauge invariant, but have been neglected in most previous works. Note that in ref. [122], all pertinent diagrams for the process $K^-p \to MB\gamma$ were drawn but not all were actually calculated.

Another simplification in chiral unitary approaches is the reduction of the interaction kernel to the on-shell point. Although the interaction kernel appears in loops, it has been argued that the off-shell components can be absorbed by redefining the coupling constants, as discussed in sec. 1.5 (see also [125]). Here we do not employ the on-shell approximation but present two alternative methods to retain the off-shell components in the interaction kernel. For a discussion of these issues, we refer the reader to ref. [60].

In chiral unitary approaches to meson photoproduction, it has been common practice to utilize s-wave projections for the meson-baryon scattering kernel and the photoproduction multipoles. We will also analyze the accuracy of this approximation since in our approach some higher partial waves are generated through the small components of the Dirac spinors describing the baryon octet fields and through the kaon (baryon) pole term in charged (neutral) meson production.

The main goal of the present chapter is the construction of a minimal approach to meson photoproduction based on the chiral effective Lagrangian which is exactly unitary and gauge invariant. The presented method fulfills these important requirements from field theory, while at the same time any subset of diagrams cannot be omitted as this would violate unitarity or gauge invariance. In this study, we restrict ourselves to the chiral effective Lagrangian at leading order. The inclusion of higher chiral orders in the interaction kernel is straightforward and necessary to obtain better agreement with experiment, particularly at higher energies away from the respective thresholds. This will be the subject of forthcoming work.

This chapter is organized as follows. In the next section, the effective Lagrangian and the Bethe-Salpeter formalism including off-shell components are introduced. The gauge invariant extension to photo- and electroproduction processes is discussed in sect. 5.3. Section 5.4 contains the comparison with experimental data on kaon photoproduction. The phenomenological impact of the violation of gauge invariance and the use of the on-shell approximation is also discussed. We summarize our findings in sec. 5.5, while lengthy formulae and a second, alternative method for including off-shell pieces in the interaction kernel are relegated to appendices E-J.

5.2 Bethe-Salpeter equation

The chiral effective Lagrangian incorporates symmetries and symmetry-breaking patterns of QCD in a model-independent way, in particular chiral symmetry and its explicit breaking through the finite quark masses. By expanding Green functions in powers of Goldstone boson masses and small momenta a chiral counting scheme can be established. However, the strict perturbative chiral expansion is only applicable at low energies, and it certainly fails in the vicinity of resonances. In this respect, the combination of the chiral effective Lagrangian with non-perturbative schemes based on coupled channels and the Bethe-Salpeter equation (BSE) have proven useful both in the purely mesonic and in the meson-baryon sector [56, 57, 104, 120, 121]. Such approaches extend the range of applicability of the chiral effective Lagrangian by implementing exact two-body unitarity in a non-perturbative fashion and generating resonances dynamically.

Here we restrict ourselves to the meson-baryon Lagrangian at leading order, given in eq. (1.33). The electromagnetic interaction is implemented in a similar way as in sec. 4.3. Here we set $v_\mu = eQA_\mu$ (so that e is negative), where $Q = \frac{1}{3}\text{diag}(2, -1, -1)$ is the quark charge matrix. We also use

$$u_\mu = iu^\dagger \nabla_\mu U u^\dagger,$$
$$\nabla_\mu U = \partial_\mu U - i[v_\mu, U].$$

The matrices u and $U = u^2$ have been defined in terms of the meson fields in sec. 1.2 and 1.3, see eq. (1.13). In our numerical work, we use the coupling constants $D = 0.8, F = 0.46$ [33] in the leading order meson-baryon Lagrangian. By expanding the chiral connection in powers of the meson fields, one derives from the effective Lagrangian the leading order $\phi^2 \bar{B} B$ vertex (the so-called 'Weinberg-Tomozawa' (WT) term), which we use as the driving term in our Bethe-Salpeter equation. Stated differently, this vertex insertion is the 'interaction kernel' of the integral equation. One finds for the corresponding potential V (which is the WT-vertex graph multiplied by i)

$$V^{bj,ai}(\slashed{q}_2, \slashed{q}_1) = g^{bj,ai}(\slashed{q}_1 + \slashed{q}_2). \tag{5.1}$$

Here, q_1 and q_2 are the four-momenta of the incoming and the outgoing meson, respectively, and the coupling constants are summarized as a matrix in channel space, with the entries

$$g^{bj,ai} = -\frac{1}{4F_j F_i} \langle \lambda^{b\dagger}[[\lambda^{j\dagger}, \lambda^i], \lambda^a]\rangle, \tag{5.2}$$

where λ^a are the generators of the $SU(3)$ Lie-Algebra in the physical (particle) basis. In this representation, a double index bj specifies a particular channel consisting of a baryon b and

Figure 5.1: Graphical illustration of the BSE for meson-baryon scattering. The filled circle represents the full scattering matrix and the open square the driving meson-baryon vertex.

a meson j. In the above expressions, ai specifies the channel of the incoming particles, while bj labels the outgoing meson-baryon state. Note that we use different values for the meson decay constants F_i (where again i labels the meson in the corresponding channel), instead of the meson decay constant in the chiral limit. In practice, these three constants F_π, F_K, F_η will be used as fit parameters, which are allowed to vary in a reasonable range, to be specified in sec. 5.4. We refer to the latter section for a discussion of this issue.

The baryon and the meson propagator, iS and $i\Delta$, are also summarized as (diagonal) matrices in channel space. They are given by

$$iS^{bj,ai}(\slashed{p}) = \frac{i\delta^{ba}\delta^{ji}}{\slashed{p} - m_a}, \qquad (5.3)$$

$$i\Delta^{bj,ai}(p) = \frac{i\delta^{ba}\delta^{ji}}{p^2 - M_j^2}. \qquad (5.4)$$

We can now write down the integral equation for the meson-baryon scattering amplitude $T^{bj,ai}$ in a rather compact form (suppressing the channel indices, but remembering the matrix character of the various amplitudes):

$$T(\slashed{q}_2, \slashed{q}_1; p) = V(\slashed{q}_2, \slashed{q}_1) + \int \frac{d^d l}{(2\pi)^d} V(\slashed{q}_2, \slashed{l}) iS(\slashed{p} - \slashed{l}) \Delta(l) T(\slashed{l}, \slashed{q}_1; p). \qquad (5.5)$$

Here, $p \equiv p_1 + q_1 = p_2 + q_2$ is the overall momentum, where p_1 and p_2 are the four momenta of the incoming and outgoing baryon, respectively. The BSE is illustrated in fig. 5.1. Note that we use dimensional regularization throughout. The solution of the BSE reads

$$T(\slashed{q}_2, \slashed{q}_1; p) = W(\slashed{q}_2, \slashed{q}_1; p) + W(\slashed{q}_2, \tilde{p}; p) G(p) [1 - W(\tilde{p}, \tilde{p}; p) G(p)]^{-1} W(\tilde{p}, \slashed{q}_1; p) \qquad (5.6)$$

with

$$W(\slashed{q}_2, \slashed{q}_1; p) = \slashed{q}_2 g \frac{1}{1 + I_M g} + \frac{1}{1 + g I_M} g \slashed{q}_1 - g \frac{1}{1 + I_M g} I_M (\slashed{p} - m) \frac{1}{1 + g I_M} g. \qquad (5.7)$$

$W(\slashed{q}_2, \tilde{p}; p)$ can be obtained from eq. (5.7) by replacing $\slashed{q}_1 \to \slashed{p} - m \equiv \tilde{p}$, and so on, m is the baryon mass matrix with entries

$$m^{bj,ai} = \delta^{ba}\delta^{ji} m_a, \qquad (5.8)$$

and the loop integrals are given by

$$I_M^{bj,ai} = \int \frac{d^d l}{(2\pi)^d} i\Delta^{bj,ai}(l), \qquad (5.9)$$

$$G^{bj,ai}(p) = \int \frac{d^d l}{(2\pi)^d} i[\Delta(l) S(\slashed{p} - \slashed{l})]^{bj,ai}. \qquad (5.10)$$

Note that in eqs. (5.6) and (5.7), the symbol '1' represents the unit matrix in channel space, while matrix-valued denominators denote matrix inversion. The proof that the above expression for T is indeed a solution of the BSE can be found in app. E.

It is important to keep in mind that all the involved expressions are matrices in channel space, so e.g. I_M does in general not commute with the coupling matrix g. The on-shell substitutions

$$\slashed{q}_{1,2} \to \slashed{p}_{1,2} + \slashed{q}_{1,2} - m = \slashed{p} - m \equiv \tilde{p}, \qquad (5.11)$$

which occur in the above expressions, provide the connection to the alternative method of solving the BSE presented in app. G. The solution given there is completely equivalent to the one described above. Here, again, we note that $\tilde{p}$ is a matrix in channel space due to the matrix character of the baryon mass matrix m.

Putting the external momenta on their respective mass shells, the form of the solution T of eq. (5.5) simplifies to

$$T_{\text{on}} = [1 - WG]^{-1} W, \qquad (5.12)$$

where we used the shorthand notation $W \equiv W(\tilde{p}, \tilde{p}; p)$ and $G \equiv G(p)$. To obtain the complete on-shell meson-baryon scattering amplitude, this expression has to be sandwiched between baryon spinors $\bar{u}(p_2)\ldots u(p_1)$. Neglecting the tadpole integrals I_M in W, one obtains from eq. (5.12)

$$T'_{\text{on}} = [1 - VG]^{-1} V, \qquad (5.13)$$

which is the form for the scattering amplitude usually encountered when utilizing the 'on-shell-approximation' (which consists of setting also internal loop momenta in the potential V on-shell) to arrive at a simplified version of the BSE. The integral equation eq. (5.5) then reduces to an algebraic equation that can be solved by matrix inversion to give an expression like T'_{on} above. Retaining the off-shell pieces in the exact solution T_{on} therefore corresponds to keeping the terms proportional to I_M in the expression for W.

In [125] it was argued that the on-shell reduction of the interaction kernel is justified as the off-shell pieces can be absorbed into the vertices. While this procedure simplifies the interaction kernel and the treatment of the BSE, it is in fact not necessary in order to solve the BSE as explained above. We choose not to discard the off-shell pieces and evaluate the diagrams in the way dictated by field theory, keeping in mind, however, that the off-shell parts are not uniquely fixed by any physical requirements and depend on the field parametrization [60]. Recall that eq. (1.13) is not the only possible choice for the parametrization of the meson fields, and that the off-shell amplitudes in ChPT depend on this choice.

Another important reason for taking the full off-shell dependence of the interaction kernel into account is that the implementation of the BSE in a gauge-invariant description of photoproduction processes is then straightforward. This is because we evaluate the occurring loop graphs according to the rules of field theory which also ensures that gauge symmetry can be incorporated in a standard way. This is not the case when utilizing the on-shell approximation. At present, our interaction kernel is restricted to the lowest order contact term derived from the effective meson-baryon Lagrangian. The extension to more general vertex structures in the kernel is possible and will be discussed in future work. For now, it is not our aim to achieve perfect agreement with experimental data (in which case higher-order terms in the kernel would be indispensable), but to construct the simplest possible amplitude for kaon photoproduction that reconciles the framework of chiral unitary approaches with the fundamental principle of gauge invariance in a straightforward way.

For the sake of completeness, we add some comments on unitarity here. From the expression for T_{on}, eq. (5.12), one immediately confirms the condition for two-particle unitarity

$$\text{Im}(T_{on}^{-1}) = -\text{Im}(G). \tag{5.14}$$

The reason that the unitarity condition can be expressed in this simple form (instead of the general form of eq. (1.41)) is given by the fact that the on-shell scattering amplitude T_{on} depends only on the overall momentum p. For the more general unitarity condition, we refer to sec. 4.5, see e.g. eq. (4.40).

Our approach to photoproduction presented in the next section will be such that unitarity is also satisfied when the photon is coupled to the meson-baryon bubble chain summed up by means of the BSE, using the method explained in the previous chapter.

The form of the solution T_{on} guarantees that the above unitarity statement holds. However, the careful reader might have noticed that the integrals I_M and $G(p)$ occurring in the amplitude are divergent for $d \to 4$ and that the introduction of appropriate counterterms might spoil the simple form of the solution given in eq. (5.12). To deal with this complication, one can show by means of a rather lengthy calculation, presented in app. F, that the modification

$$V \to V + \delta V \equiv V_\delta, \tag{5.15}$$

of the interaction kernel V in the BSE, where

$$V_\delta(\slashed{q}_2, \slashed{q}_1; p) = W_\delta(\slashed{q}_2, \slashed{q}_1; p) - W_\delta(\slashed{q}_2, \tilde{p}; p)\delta G(p)[1 + W_\delta(\tilde{p}, \tilde{p}; p)\delta G(p)]^{-1} W_\delta(\tilde{p}, \slashed{q}_1; p) \tag{5.16}$$

and

$$W_\delta(\slashed{q}_2, \slashed{q}_1; p) = \slashed{q}_2 g \frac{1}{1 - \delta I_M g} + \frac{1}{1 - g\delta I_M} g\slashed{q}_1 + g \frac{1}{1 - \delta I_M g} \delta I_M (\slashed{p} - m) \frac{1}{1 - g\delta I_M} g \tag{5.17}$$

leads to a new integral equation with a solution T_δ. The solution T_δ is obtained from T by replacing

$$G \to G - \delta G$$
$$I_M \to I_M - \delta I_M$$

in eqs. (5.6) and (5.7). One observes that V_δ has the same form as T, with G and I_M replaced by $-\delta G$ and $-\delta I_M$, respectively. The corrections δG and δI_M have the form of polynomial terms which serve to absorb the divergences in the loop integrals. The divergences have thus been shifted from the loop integrals in T to the new kernel. Of course, this is not a renormalization scheme in the usual sense, since the additional terms δV in V_δ do not correspond exactly to counterterms derived from an effective Lagrangian which is obvious from the lack of crossing symmetry of the amplitude. It has already been noted in ref. [60] that the solution of the BSE can not be renormalized within a usual renormalization scheme. However, the foregoing discussion shows at least that altering the loop integrals appearing in the solution of the BSE by terms δG and δI_M is equivalent to certain modifications of the potential. In practice, we merely omit the divergences in the loop integrals in our expression for T, which will therefore depend on the regularization scale μ showing up in the modified loop integrals.

This form of the off-shell BSE solution T is not particularly convenient for the purpose of implementing it into the photoproduction amplitude. Therefore, we rewrite it by using the following decompositions:

$$G(p) = G_1(p)\slashed{p} + G_0(p), \tag{5.18}$$
$$W(\tilde{p},\tilde{p};p) = W_1(p)\slashed{p} + W_0(p), \tag{5.19}$$

where the matrices $G_{0,1}$ follow straightforwardly from the explicit expression for the loop integral G which is given in eq. (H.11) of app. H, and

$$W_1(p) = g\frac{2 + I_M g}{[1 + I_M g]^2}, \tag{5.20}$$

$$W_0(p) = g\frac{1}{1 + I_M g}(I_M m)\frac{1}{1 + gI_M}g - mg\frac{1}{1 + I_M g} - \frac{1}{1 + gI_M}gm. \tag{5.21}$$

Note that $G_i(p)$ and $W_i(p)$ are Lorentz scalars which depend only on the variable p^2. Using these expressions one can derive

$$1 - W(\tilde{p},\tilde{p};p)G(p) = \tilde{W}_1\slashed{p} + \tilde{W}_0, \tag{5.22}$$

with

$$\tilde{W}_1 = -(W_1(p)G_0(p) + W_0(p)G_1(p)), \tag{5.23}$$
$$\tilde{W}_0 = 1 - p^2 W_1(p)G_1(p) - W_0(p)G_0(p). \tag{5.24}$$

The above results yield

$$G(p)[1 - W(\tilde{p},\tilde{p};p)G(p)]^{-1} = \Omega_1(p)\slashed{p} + \Omega_0(p) \tag{5.25}$$

with

$$\Omega_1(p) = G_0(p)[p^2\tilde{W}_1 - \tilde{W}_0\tilde{W}_1^{-1}\tilde{W}_0]^{-1}$$
$$\qquad - G_1(p)\tilde{W}_1^{-1}\tilde{W}_0[p^2\tilde{W}_1 - \tilde{W}_0\tilde{W}_1^{-1}\tilde{W}_0]^{-1}, \tag{5.26}$$
$$\Omega_0(p) = p^2 G_1(p)[p^2\tilde{W}_1 - \tilde{W}_0\tilde{W}_1^{-1}\tilde{W}_0]^{-1}$$
$$\qquad - G_0(p)\tilde{W}_1^{-1}\tilde{W}_0[p^2\tilde{W}_1 - \tilde{W}_0\tilde{W}_1^{-1}\tilde{W}_0]^{-1}. \tag{5.27}$$

All these results can be combined to provide the decomposition of the off-shell BSE-solution into the following independent Clifford algebra structures:

$$T(\slashed{q}_2,\slashed{q}_1;p) = \slashed{q}_2\slashed{p}\slashed{q}_1 T_1(p) + \slashed{q}_2\slashed{q}_1 T_2(p) + \slashed{p}\slashed{q}_1 T_3(p) + \slashed{q}_2\slashed{p} T_4(p) + \slashed{q}_1 T_5(p) + \slashed{q}_2 T_6(p) + \slashed{p} T_7(p) + T_8(p). \tag{5.28}$$

The scalar coefficient functions $T_n(p)$ read

$$\begin{aligned}
T_1(p) &= L_1\Omega_1(p)L_1, \\
T_2(p) &= L_1\Omega_0(p)L_1, \\
T_3(p) &= [L_2\Omega_0(p) + L_3\Omega_1(p)]L_1, \\
T_4(p) &= T_3^T(p), \\
T_5(p) &= [p^2 L_2\Omega_1(p) + L_3\Omega_0(p)]L_1 + L_1, \\
T_6(p) &= T_5^T(p), \\
T_7(p) &= [p^2 L_2\Omega_1(p) + L_3\Omega_0(p)]L_2 \\
&\quad + [L_2\Omega_0(p) + L_3\Omega_1(p)]L_3^T - gI_M L_2, \\
T_8(p) &= p^2[L_2\Omega_0(p)L_2 + L_3\Omega_1(p)L_2 + L_2\Omega_1(p)L_3^T] \\
&\quad + L_3\Omega_0(p)L_3^T - L_3 I_M g,
\end{aligned} \tag{5.29}$$

where the superscript T denotes transposition of channel indices and

$$L_1 = g\frac{1}{1+I_M g},$$

$$L_2 = \frac{1}{[1+gI_M]^2}g,$$

$$L_3 = -\frac{1}{1+gI_M}gm\frac{1}{1+I_M g}.$$

Note that $L_1^T = L_1$ and $L_2^T = L_2$, but in general $L_3^T \neq L_3$. For external on-shell particles, we can make the usual substitutions $\slashed{q}_1, \slashed{q}_2 \to \slashed{p} - m \equiv \tilde{p}$ to arrive at the on-shell scattering amplitude

$$T_{\text{on}} = T_{\text{on}}^{(1)}\slashed{p} + T_{\text{on}}^{(0)} \tag{5.30}$$

with

$$\begin{aligned}
T_{\text{on}}^{(1)} &= p^2 T_1 + mT_1 m - mT_2 - T_2 m \\
&\quad - T_3 m - mT_4 + T_5 + T_6 + T_7,
\end{aligned} \tag{5.31}$$

$$\begin{aligned}
T_{\text{on}}^{(0)} &= p^2(T_2 + T_3 + T_4 - T_1 m - mT_1) + mT_2 m \\
&\quad - T_5 m - mT_6 + T_8.
\end{aligned} \tag{5.32}$$

The functions T_n given in eq. (5.29) depend only on the variable p^2. They will enter the calculation of the various photo- and electroproduction amplitudes in the next section.

5.3 Photo- and electroproduction

The Bethe-Salpeter approach discussed in the last section (or the Lippmann-Schwinger equation in the non-relativistic framework) can be implemented in electroproduction processes of mesons on the nucleon. In previous work, the electromagnetic meson production on the nucleon was calculated at tree level and the produced meson-baryon pair was subject to final-state interactions [56,91]. As the photon does not couple to all intermediate states of the meson-baryon bubble chain, gauge invariance is in general violated and must be restored via artificial manipulations. Therefore it seems desirable to develop a formalism which implements the principles of gauge invariance and unitarity in a most natural and straightforward manner. We follow here the path which we have already outlined in the previous chapter, in rather general terms, for the case of a photon coupling to a meson-baryon scattering amplitude. In this work, we shall be more explicit in evaluating the contributions to the various amplitudes in question. Our approach for constructing a unitary and gauge invariant electroproduction amplitude decomposes into two major steps:

(1) Fix the hadronic part of the amplitude by making use of a BSE to implement exact two-body unitarity.

(2) Couple the photon to the 'hadronic skeleton' constructed in step (1) wherever possible, i.e. to all external and internal lines describing the propagation of the involved particles as well as to (momentum-dependent) vertices.

The procedure of step (2), which is the most natural way to guarantee gauge invariance of the amplitude, leads to contributions that were usually not considered in chiral unitary approaches involving electromagnetic interactions. The importance of these additional contributions which render the electroproduction amplitude gauge invariant can also be quantified within the scheme utilized here. The coupling of the photon to internal lines in the bubble chain generated by the BSE leads to rather involved expressions, since the meson-baryon scattering amplitude T appears twice in the corresponding amplitudes, as will be made more explicit later when we give the formal expressions for the various contributions.

Exact unitarity (in the subspace of meson-baryon states) is a fundamental principle satisfied by our approach. It is important to note that the procedure of step (2) above does not spoil the requirement of unitarity we built in by means of the BSE in step (1), as we will show below. First, we have to specify the set of graphs which constitute our amplitude.

We start with the tree level $\bar{B}\phi B$ amplitude derived from the leading order Lagrangian of eq. (1.33),

$$\hat{V}^{bi,a} = \slashed{q}\,\gamma_5\,\hat{g}^{bi,a}\,. \tag{5.33}$$

The last index a labels the incoming baryon (here always the proton), while the double index belongs to the outgoing pair of baryon b and meson i. For fixed a, $\hat{V}$ is a vector in channel space. Furthermore, q is the four-momentum of the outgoing meson and

$$\hat{g}^{bi,a} = -\frac{D}{\sqrt{2}F_i}\langle\lambda^{b\dagger}\{\lambda^{i\dagger},\lambda^a\}\rangle - \frac{F}{\sqrt{2}F_i}\langle\lambda^{b\dagger}[\lambda^{i\dagger},\lambda^a]\rangle\,. \tag{5.34}$$

To this tree level amplitude, we add the loop contribution that accounts for the final-state interaction after the meson has left the vertex, to obtain (cf. fig. 5.2)

$$\Gamma(\slashed{q},\slashed{p}) = \slashed{q}\gamma_5\hat{g} + \int\frac{d^d l}{(2\pi)^d}T(\slashed{q},l;p)iS(\slashed{p}-\slashed{l})\Delta(l)\slashed{l}\gamma_5\hat{g}\,. \tag{5.35}$$

We call this approximation to the full meson-baryon interaction the 'turtle approximation', because with some imagination the right-most diagram in fig. 5.2 resembles a turtle. Here and

Figure 5.2: The dressed meson-baryon vertex in the turtle approximation. The filled circle (open square) denotes the full (tree level) meson-baryon interaction.

in the following we again suppress the channel indices for brevity, and p is the momentum of the incoming baryon. Using the explicit form of T given in eq. (5.28), we find

$$\Gamma(\not{q},\not{p}) = \not{q}\gamma_5 \hat{g} + T(\not{q},\not{p}-m;p)[(\not{p}-m)G(p) - I_{\mathrm{M}}]\gamma_5 \hat{g}$$
$$= [\not{q}\not{p}\Gamma_1(p) + \not{q}\Gamma_2(p) + \not{p}\Gamma_3(p) + \Gamma_4(p)]\gamma_5 \qquad (5.36)$$

with scalar coefficient functions Γ_n given by

$$\Gamma_1(p) = T_1(p^2 H_1 - mH_0) + T_2(H_0 - mH_1) + T_4 H_0 + T_6 H_1 , \qquad (5.37)$$
$$\Gamma_2(p) = \hat{g} + p^2 T_1(H_0 - mH_1) + T_2(p^2 H_1 - mH_0) + p^2 T_4 H_1 + T_6 H_0 , \qquad (5.38)$$
$$\Gamma_3(p) = T_3(p^2 H_1 - mH_0) + T_5(H_0 - mH_1) + T_7 H_0 + T_8 H_1 , \qquad (5.39)$$
$$\Gamma_4(p) = p^2 T_3(H_0 - mH_1) + T_5(p^2 H_1 - mH_0) + p^2 T_7 H_1 + T_8 H_0 , \qquad (5.40)$$

where

$$H_1 = (G_0(p) - mG_1(p))\hat{g} ,$$
$$H_0 = (p^2 G_1(p) - mG_0(p) - I_{\mathrm{M}})\hat{g} .$$

The functions $G_{0,1}(p)$ have been defined in eq. (5.18).

In order to complete step (1), we still have to specify which meson-baryon channels contribute in the framework of our electroproduction model. In this first study, we choose to consider only the ground-state octets of mesons and baryons, respectively. Moreover, from the topology of the hadronic part of the Feynman graph in fig. 5.2 we can conclude that the meson-baryon pairs must have the charge and strangeness quantum numbers of the proton. This limits the number of channels to six:

$$p\pi^0,\; n\pi^+,\; p\eta,\; \Lambda K^+,\; \Sigma^0 K^+,\; \Sigma^+ K^0 . \qquad (5.41)$$

The limitation to these channels can only occur because our amplitude is not crossing-symmetric, otherwise more channels with different quantum numbers must be considered. The violation of crossing-symmetry is a drawback of the BSE method which we use to iterate the rescattering graphs. To our knowledge, the implementation of both crossing symmetry and exact (two-body) unitarity on the basis of Feynman diagrams has not yet been accomplished. Other methods such as, e.g., an analog of the Roy equations (or generalizations thereof) for pion-pion, pion-kaon or pion-nucleon scattering, see [127–129], might be required to achieve this. Here, our goal is more modest, and we sacrifice crossing symmetry in favor of unitarity.

By now, we have finished the first part (step (1)) of our program. Our next task is to couple the photon to the hadronic part of the amplitude in a gauge-invariant fashion. Inserting the photon coupling at every possible place leads to the set of diagrams displayed in fig. 5.3.

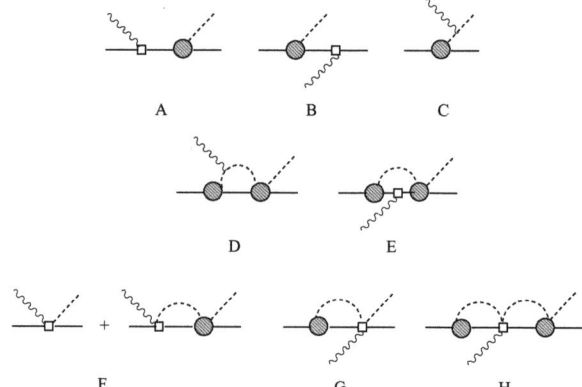

Figure 5.3: Classes of diagrams for kaon production off the nucleon in the turtle approximation utilized here.

The first three graphs are generated by coupling the photon to the external lines of the hadronic part. This leads to the expressions

$$S_s^\mu = \Gamma(\slashed{q},\slashed{p})iS(\slashed{p})(ie\gamma^\mu),\tag{5.42}$$
$$S_u^\mu = (ieQ_B\gamma^\mu)iS(\slashed{p}_1 - \slashed{q})\Gamma(\slashed{q},\slashed{p}_1),\tag{5.43}$$
$$S_t^\mu = ieQ_M(2q-k)^\mu i\Delta(q-k)\Gamma(\slashed{q}-\slashed{k},\slashed{p}_1).\tag{5.44}$$

Here and in the following, k is the four-momentum of the incoming photon, q is the four-momentum of the outgoing Goldstone boson and p_1 and p_2 are the four-momenta of the incoming and outgoing baryon, respectively, while $p \equiv p_1 + k$. We have also introduced the meson (M) and baryon (B) charge matrices. They are diagonal matrices in channel space and read

$$Q_M = \text{diag}(0,1,0,1,1,0),$$
$$Q_B = \text{diag}(1,0,1,0,0,1).$$

The expressions for the three amplitudes $S^\mu_{s,u,t}$ must be decomposed into their independent Lorentz structures which is outlined in app. I. Next we consider the coupling of the photon to internal lines of the bubble chain, leading to those diagrams that are most tedious to work out. Consider first the coupling of the photon to an internal baryon line, see fig. 5.3 E. Using the pertinent Feynman rules, the contribution of this graph is easily seen to be

$$S_B^\mu = -i\int\frac{d^d l}{(2\pi)^d}T(\slashed{q},\slashed{l};p)S(\slashed{p}-\slashed{l})\Delta(l)eQ_B\gamma^\mu S(\slashed{p}_1-\slashed{l})\Gamma(\slashed{l},\slashed{p}_1),\tag{5.45}$$

while the coupling of the photon to an internal meson line (fig. 5.3 D) leads to

$$S_M^\mu = -i\int\frac{d^d l}{(2\pi)^d}T(\slashed{q},\slashed{p}-\slashed{l};p)S(\slashed{l})\Delta(p-l)eQ_M(2(p_1-l)+k)^\mu\Delta(p_1-l)\Gamma(\slashed{p}_1-\slashed{l},\slashed{p}_1).\tag{5.46}$$

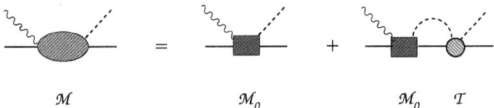

Figure 5.4: Illustration of the integral equation (5.52) satisfied by our electroproduction amplitude

Again, the decomposition into independent Lorentz structures can be found in app. I.
The next class of graphs we consider arises due to the 'Kroll-Ruderman' (KR) term stemming from the covariant derivative $\nabla_\mu U$ in the chiral Lagrangian. The corresponding tree level vertex reads (in matrix form)

$$S^\mu_{\text{KR,tree}} = eQ_M \hat{g}\, \gamma^\mu \gamma_5 \,. \tag{5.47}$$

We add the loop contribution that accounts for the final-state interaction after the meson has left the KR vertex, to arrive at (see fig. 5.3 F)

$$S^\mu_{\text{KR}} = S^\mu_{\text{KR,tree}} + \int \frac{d^d l}{(2\pi)^d} T(\slashed{q}, \slashed{l}; p) i S(\slashed{p} - \slashed{l}) \Delta(l) S^\mu_{\text{KR,tree}} \,. \tag{5.48}$$

The remaining graphs are depicted in figs. 5.3 G,H. These contributions arise from the terms proportional to the external vector field in the chiral connection Γ^μ, leading to a $\bar{B}\phi\phi\gamma B$ vertex rule

$$-ie\,\gamma^\mu \{Q_M, g\} \,, \tag{5.49}$$

and to the following expressions for the Feynman graphs of fig. 5.3 G,H:

$$S^\mu_{\text{WT1}} = e\gamma^\mu \{Q_M, g\} \int \frac{d^d l}{(2\pi)^d} i S(\slashed{p}_1 - \slashed{l}) \Delta(l) \Gamma(\slashed{l}, \slashed{p}_1) \,, \tag{5.50}$$

$$S^\mu_{\text{WT2}} = \int \frac{d^d \tilde{l}}{(2\pi)^d} T(\slashed{q}, \slashed{\tilde{l}}; p) i S(\slashed{p} - \slashed{\tilde{l}}) \Delta(\tilde{l}) S^\mu_{\text{WT1}} \,. \tag{5.51}$$

The Lorentz structure decomposition of all the above diagrams and the according contributions to the invariant amplitudes B_i are given in app. I.
Having finished the construction of the electroproduction amplitude, we return to the issue of unitarity. The crucial observation here is that every electroproduction amplitude $\mathcal{M}^\mu(q, k; p)$ which may be written as (see also fig. 5.4)

$$\mathcal{M}^\mu(q, k; p) = \mathcal{M}^\mu_0(q, k; p) + \int \frac{d^d l}{(2\pi)^d} T(q, l; p) i S(\slashed{p} - \slashed{l}) \Delta(l) \mathcal{M}^\mu_0(l, k; p) \tag{5.52}$$

obeys the requirement of two-body unitarity in the subspace of meson-baryon channels. Here, T is an amplitude for meson-baryon scattering that solves a BSE of the type of eq. (5.5) and $\mathcal{M}^\mu_0$ is a real kernel.
The proof proceeds in close analogy to the one presented in sec. 4.5 of chapter 4. Now we note that our electroproduction amplitude given above decomposes into five 'unitarity classes', each

of which obey an integral equation of the above type (5.52) by construction (with $\mathcal{T} = T$):

$$\text{Class 1} : S_s^\mu$$
$$\text{Class 2} : S_u^\mu + S_B^\mu$$
$$\text{Class 3} : S_t^\mu + S_M^\mu$$
$$\text{Class 4} : S_{KR}^\mu$$
$$\text{Class 5} : S_{WT1}^\mu + S_{WT2}^\mu.$$

Hence, each class leads to a electroproduction amplitude that obeys unitarity for itself, though not gauge invariance. This is because the expressions S_{WT1}^μ, S_t^μ and S_u^μ as well as the tree graphs of S_{KR}^μ and S_s^μ are real in the physical region for the electroproduction process and can therefore constitute a kernel $\mathcal{M}_0^\mu$ in eq. (5.52). The sum

$$\mathcal{M}^\mu \equiv S_s^\mu + S_u^\mu + S_t^\mu + S_B^\mu + S_M^\mu + S_{KR}^\mu + S_{WT1}^\mu + S_{WT2}^\mu \tag{5.53}$$

will then also obey unitarity, due to the linearity of the integral equation (5.52) in $\mathcal{M}_0^\mu$.

We will now show that the sum $\mathcal{M}^\mu$ of the five classes is gauge invariant by proving $k \cdot \mathcal{M} = 0$ for on-shell mesons and baryons. This might be obvious from our construction, but we include the proof for completeness. The contraction of k with the different amplitudes yields

$$k \cdot S_s = -e\Gamma(\slashed{q}, \slashed{p}),$$
$$k \cdot S_u = eQ_B\Gamma(\slashed{q}, \slashed{p}_1),$$
$$k \cdot S_t = eQ_M\Gamma(\slashed{q} - \slashed{k}, \slashed{p}_1),$$
$$k \cdot S_B = \int \frac{d^d l}{(2\pi)^d} T(\slashed{q}, \slashed{l}; p) iS(\slashed{p} - \slashed{l}) \Delta(l) eQ_B \Gamma(\slashed{l}, \slashed{p}_1)$$
$$- \int \frac{d^d l}{(2\pi)^d} T(\slashed{q}, \slashed{l}; p) eQ_B iS(\slashed{p}_1 - \slashed{l}) \Delta(l) \Gamma(\slashed{l}, \slashed{p}_1),$$
$$k \cdot S_M = \int \frac{d^d l}{(2\pi)^d} T(\slashed{q}, \slashed{l}; p) iS(\slashed{p} - \slashed{l}) \Delta(l) eQ_M \Gamma(\slashed{l} - \slashed{k}, \slashed{p}_1)$$
$$- \int \frac{d^d l}{(2\pi)^d} T(\slashed{q}, \slashed{l} + \slashed{k}; p) eQ_M iS(\slashed{p}_1 - \slashed{l}) \Delta(l) \Gamma(\slashed{l}, \slashed{p}_1),$$
$$k \cdot S_{KR} = eQ_M \hat{g} \slashed{k} \gamma_5 + \int \frac{d^d l}{(2\pi)^d} T(\slashed{q}, \slashed{l}; p) iS(\slashed{p} - \slashed{l}) \Delta(l) eQ_M \hat{g} \slashed{k} \gamma_5,$$
$$k \cdot S_{WT1} = e\slashed{k}\{Q_M, g\} \int \frac{d^d l}{(2\pi)^d} iS(\slashed{p}_1 - \slashed{l}) \Delta(l) \Gamma(\slashed{l}, \slashed{p}_1),$$
$$k \cdot S_{WT2} = \int \frac{d^d \tilde{l}}{(2\pi)^d} T(\slashed{q}, \slashed{\tilde{l}}; p) iS(\slashed{p} - \slashed{\tilde{l}}) \Delta(\tilde{l}) e\slashed{k}\{Q_M, g\} \int \frac{d^d l}{(2\pi)^d} iS(\slashed{p}_1 - \slashed{l}) \Delta(l) \Gamma(\slashed{l}, \slashed{p}_1).$$

Here, charged external particles are put on-shell. First, we note that the tree graphs are gauge invariant for themselves, since the tree part of $k \cdot (S_s + S_u + S_t)$ equals

$$-e\hat{g}\slashed{q}\gamma_5 + eQ_B\hat{g}\slashed{q}\gamma_5 + eQ_M\hat{g}(\slashed{q} - \slashed{k})\gamma_5,$$

which is exactly canceled by the first term of $k \cdot S_{KR}$ (recall $Q_B + Q_M = 1$). In order to deal with the loop contributions, it is useful to rewrite the integral equations (5.5) and (5.35) for T

and Γ, respectively, as

$$T(\not{q}, \not{l}; p) = g(\not{q} + \not{l}) + \int \frac{d^d \tilde{l}}{(2\pi)^d} T(\not{q}, \tilde{\not{l}}; p) i S(\not{p} - \tilde{\not{l}}) \Delta(\tilde{l}) g(\tilde{\not{l}} + \not{l}), \tag{5.54}$$

and

$$\Gamma(\tilde{\not{l}}, \not{p}_1) = \hat{g} \tilde{\not{l}} \gamma_5 + \int \frac{d^d l}{(2\pi)^d} g(\tilde{\not{l}} + \not{l}) i S(\not{p}_1 - \not{l}) \Delta(l) \Gamma(l, \not{p}_1). \tag{5.55}$$

These equations are equivalent to eqs. (5.5) and (5.35). Using eq. (5.54) in the second term of $k \cdot S_\text{B}$ and $k \cdot S_\text{M}$, and eq. (5.55) in the first term of $k \cdot S_\text{B}$ and $k \cdot S_\text{M}$, it is straightforward to show that the loop contributions in $k \cdot \mathcal{M}$ cancel. This, together with the above result for the tree contributions, completes the proof of gauge invariance for the full amplitude $\mathcal{M}$.

5.4 Results

In this section, we will perform an overall χ^2 fit to available photoproduction and pion-induced data on the proton near the respective thresholds. In more detail, we fit the differential cross sections for photoproduction on the proton into the $K^+\Lambda$, $K^+\Sigma^0$, $K^0\Sigma^+$ final states as well as of $\pi^-p \to K^0\Lambda$, $K^0\Sigma^0$. Inspection of the differential cross sections reveals that already at moderate energies away from threshold p- and d-waves become increasingly important. Since our approach, which is based on the Weinberg-Tomozawa interaction kernel, generates mainly s-waves (with one important exception discussed below) and thus does not provide a realistic description for higher partial waves, we expect it to be valid only in the near-threshold regions. We have thus restricted our fits to energy values for which the differential cross sections are dominated by s-waves, i.e. center-of-mass energies of about 1.80 GeV corresponding to photon lab momenta of about 1.25 GeV or pion lab momenta of about 1.23 GeV. Still, we will be able to extract interesting information from such investigations, in particular, we can study in detail the commonly appearing approximations made in the literature as already mentioned in the introduction. The extension to higher energies requires inclusion of higher order counter terms from the effective Lagrangian in the derivation of the interaction kernel. However, this is beyond the scope of the present work. In this investigation, we focus on the importance of gauge invariance and study the on-shell approximation in the driving terms and are satisfied with a moderate description of both photoproduction and pion-induced data on the proton near threshold. Our investigation sets the stage for systematic improvements in the future which will lead to better agreement with experiment.

The free parameters in our approach are, on the one hand, the three meson decay constants F_π, F_K, F_η which we vary separately within realistic bounds, as the $SU(3)$ symmetry-breaking differences between them are beyond our working precision of the effective potential. More precisely, $SU(3)$ symmetry breaking is generated by various higher order terms in the meson-baryon (meson) Lagrangian starting at chiral order two (four). Since these contributions are not included in the leading order WT kernel, we simulate such effects by allowing variations in the various meson decays constants. This, of course, will no longer be done when the higher order terms in the interaction kernel have been included. One observes that the fitted decay constants tend to larger values reducing the strength of the Weinberg-Tomozawa interaction. This is consistent with findings in earlier chiral unitary studies, i.e. the WT interaction in many cases produces too strong s-waves, see e.g. [91]. We allow F_π, F_K, and F_η to vary between

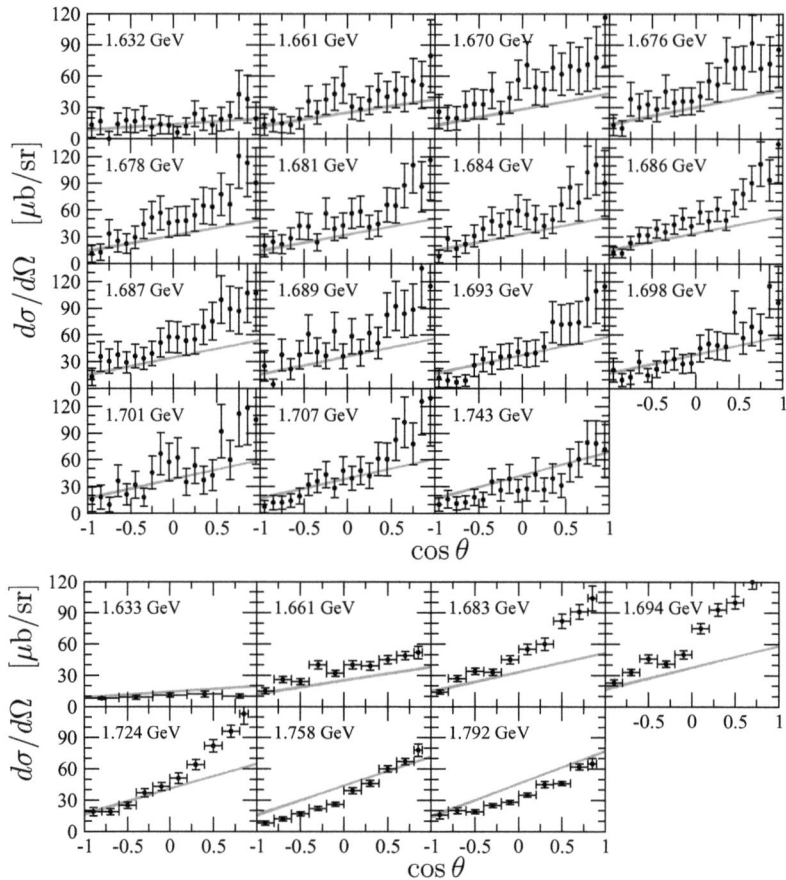

Figure 5.5: Differential cross sections for $\pi^- p \to K^0 \Lambda$ compared to data from [130] (upper panel) and [131] (lower panel). The number in each plot denotes the respective c.m. energy $\sqrt{s}$.

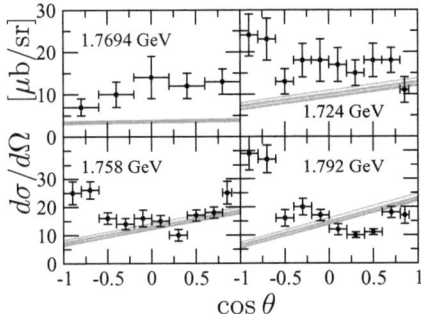

Figure 5.6: Differential cross sections for $\pi^- p \to K^0 \Sigma^0$ compared to data from [133]. The number in each plot denotes the respective c.m. energy $\sqrt{s}$.

70 MeV and 150 MeV. The values we find for the best fits, i.e. the fits with the lowest overall χ^2 values, are

$$F_\pi = (113\ldots 127)\,\text{MeV}\,,$$
$$F_K = (149\ldots 150)\,\text{MeV}\,,$$
$$F_\eta = (74\ldots 82)\,\text{MeV}\,. \qquad (5.56)$$

Note that F_K tends towards the upper limit of 150 MeV. On the other hand, we fit the four different isospin-symmetric scales μ in the loop integrals. For the best fits their values are:

$$\mu_{\pi N} = (0.46\ldots 0.54)\,\text{GeV}\,,$$
$$\mu_{\eta N} = (3.29\ldots 4.41)\,\text{GeV}\,,$$
$$\mu_{K\Lambda} = (2.56\ldots 2.86)\,\text{GeV}\,,$$
$$\mu_{K\Sigma} = (3.66\ldots 4.31)\,\text{GeV}\,. \qquad (5.57)$$

We note that these values are roughly in accordance with the natural size estimate of ref. [57] (although most of them turn out to be somewhat large).
In figs. 5.5, 5.6 we present the best fit results for the pion-induced differential cross sections $\pi^- p \to K^0 \Lambda$, $K^0 \Sigma^0$, respectively. Although the WT interaction kernel is entirely s-wave, there are p-wave contributions due to the lower components of the Dirac spinors which are introduced in the calculation of the scattering matrix T. The corresponding total cross sections and the one for $\pi^- p \to K^+ \Sigma^-$ are shown in fig. 5.7. We remark that the bands in these figures are generated by about 20 fits with a χ^2 very close to its minimal value. Since with the simple WT interaction we are not able to describe all these and the photon-induced data to a high accuracy, we refrain from giving one-sigma error bands as done in our study on $K^- p$ scattering [132]. This will be done in future work when the higher order terms in the interaction kernel will be included.
The processes $\pi^- p \to K^+ \Sigma^-$ and $\pi^+ p \to K^+ \Sigma^+$ are dominated by p- and d-waves already close to threshold. Hence, a realistic description of these channels cannot be provided by the leading WT interaction. In order not to overestimate the data we have still included the total

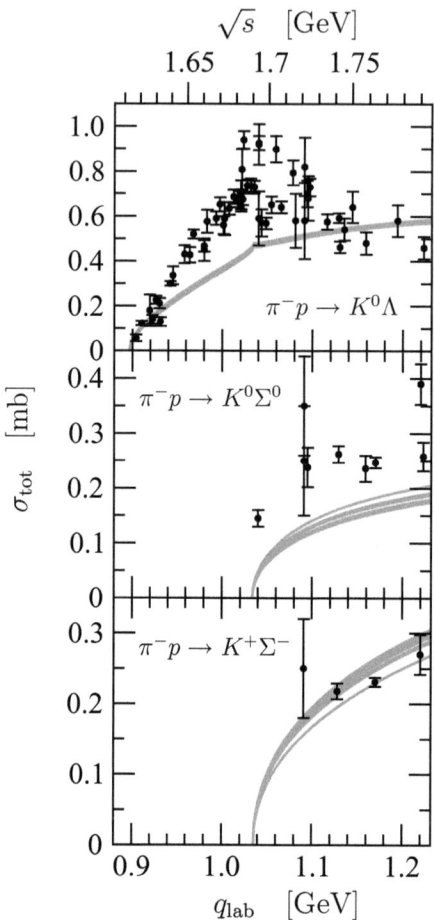

Figure 5.7: Total cross sections for π^- induced production of $K^0\Lambda$ (top), $K^0\Sigma^0$ (middle), $K^+\Sigma^-$ (bottom) as a function of the pion lab momentum q_{lab}. The corresponding c.m. energy $\sqrt{s}$ is also given. The data are taken from [130, 131, 133] and the compilation [134].

cross section for $\pi^- p \to K^+ \Sigma^-$ in our fit. We refrain, however, from taking into account data on $\pi^+ p \to K^+ \Sigma^+$ at all since, regardless of the choice of parameters (within realistic ranges), our approach clearly overshoots the $\pi^+ p \to K^+ \Sigma^+$ cross section. This indicates that higher order contact interactions in the interaction kernel are absolutely necessary—particularly for this process—in order to reduce the strength of the WT term.

Moreover, it is worthwhile mentioning that the presented fits exhibit a peak in the $\pi^- p \to \eta n$ cross section due to the $S_{11}(1535)$ resonance. Pion- and photon-induced eta production data will be included in future work after the interaction kernel has been developed to higher chiral orders.

Furthermore, we have included in our fit data on differential cross sections of the photoproduction processes $\gamma p \to K^+ \Lambda$, $K^+ \Sigma^0$, $K^0 \Sigma^+$. The results are displayed in figs. 5.8, 5.9, 5.10, respectively, while the corresponding total cross sections are presented[1] in fig. 5.11. A few remarks are in order: The SAPHIR and CLAS data on charged kaon photoproduction show some inconsistencies at forward angles, but this can not be resolved within the approximations made here. Also, the very different shape of the differential cross sections for the $K^+ \Lambda$ and $K^+ \Sigma^0$ final states can be traced back to the isospin selectivity of the Λ, see also ref. [135].

To summarize, the results presented here show a reasonable agreement with data on photon- and pion-induced reactions close to threshold, but more realistic interaction kernels and higher partial waves are required to obtain better agreement with data, in particular away from threshold. This is however beyond the scope of the present investigation.

Of interest here are gauge invariance violations encountered in previous chiral unitary approaches which only took a subset of the diagrams in fig. 5.3 into account. In order to estimate the typical size of gauge symmetry violations, we omit all contributions where the photon couples to intermediate and final states and retain only those diagrams where the photon is initially absorbed (figs. 5.3 A+F). Note that both the graphs A and F as well as their sum are unitary as explained in detail in sect. 5.3. For making this comparison, we do not refit the parameters but use the values obtained in the full approach. Some sample results are compared to the full approach in fig. 5.12. One clearly observes that violations of gauge invariance are sizable, although this effect could, in principle, be concealed numerically by readjusting the parameters of the approach. This indicates that the additional contributions which render the photoproduction amplitudes gauge invariant and were omitted in previous work are not negligible and must be taken into account.

In order to be able to compare our results with previous chiral unitary approaches we have also worked in the approximations employed in these investigations, see e.g. [91][2]. To this aim, the interaction kernel for meson-baryon scattering is directly sandwiched between Dirac spinors and projected onto the s-wave which is then iterated to infinite order in a geometric series. Furthermore, for the photoproduction process only the leading s-wave (the so-called E_{0+} multipole) is considered. Obviously, this scheme produces pure s-waves and cannot reproduce the structures from higher partial waves in the differential cross sections, see fig. 5.13. One notes that these approximations can indeed be sizable. Of course, in ref. [91] higher order terms were included in the E_{0+} multipole, but that does not change the observations just made. Again, by a suitable parameter refitting one might be able to describe the total cross sections, but given the more sophisticated scheme developed here, such approximations are no longer necessary.

---

[1] The photon lab momentum is given by $k_{\text{lab}} = (s - m_p^2)/(2m_p)$, where m_p is the proton mass.
[2] Note, however, that in ref. [91] the primary goal was the consistent inclusion of the η' meson in chiral unitary approaches.

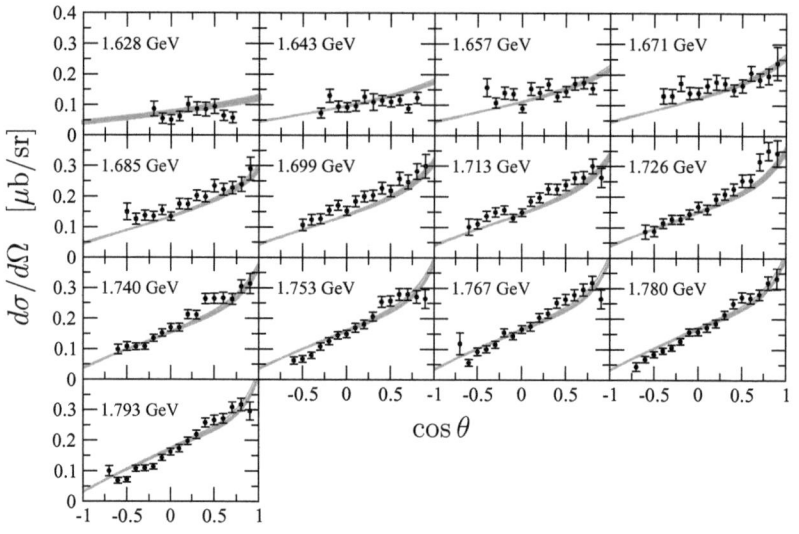
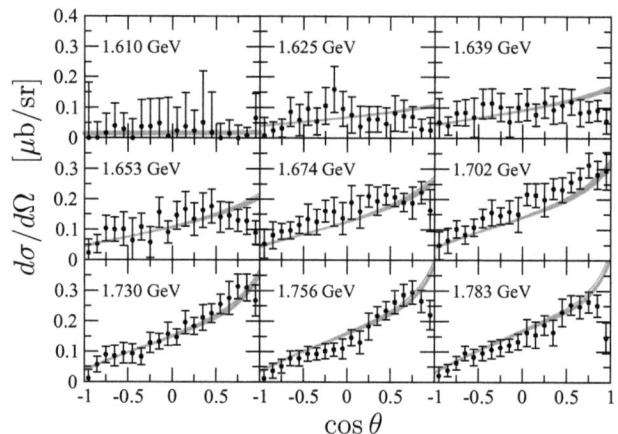

Figure 5.8: Differential cross sections for $\gamma p \to K^+\Lambda$ compared to data from [112] (upper panel) and [109] (lower panel). The number in each plot denotes the respective c.m. energy $\sqrt{s}$.

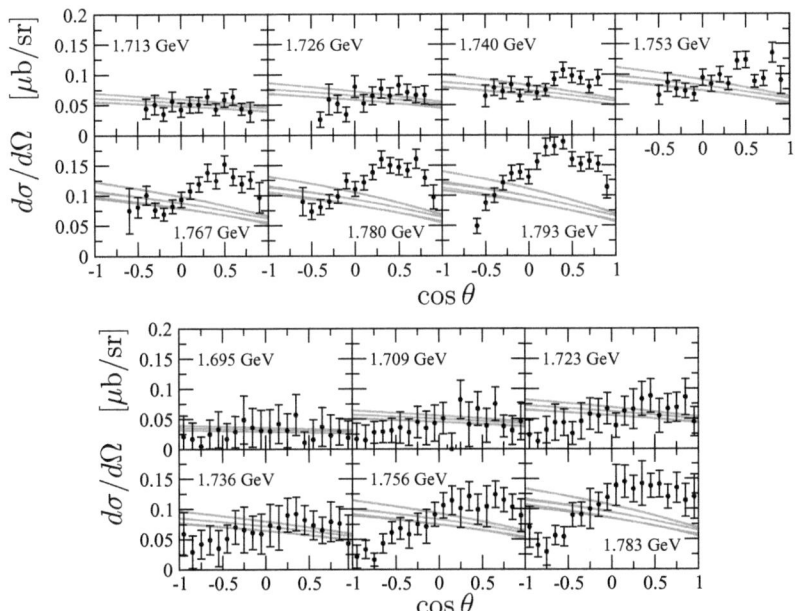

Figure 5.9: Differential cross sections for $\gamma p \to K^+ \Sigma^0$ compared to data from [112] (upper panel) and [109] (lower panel). The number in each plot denotes the respective c.m. energy $\sqrt{s}$.

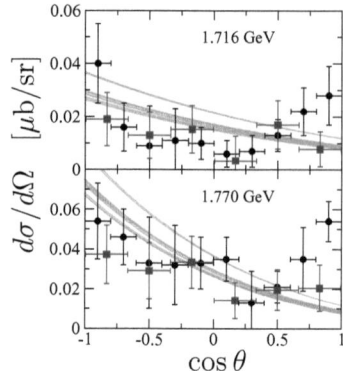

Figure 5.10: Differential cross sections for $\gamma p \to K^0\Sigma^+$ compared to data from [110] (circles) and [113] (squares). The number in each plot denotes the respective c.m. energy $\sqrt{s}$.

Overall, we have illustrated that both the on-shell approximation and the omission of certain classes of diagrams which are required to fulfill gauge invariance constitute crude approximations utilized within chiral unitary approaches in the past. Although these effects can numerically be concealed to a large extent by readjusting the parameters, important theoretical constraints are not met in this manner. On the other hand, the approach we have developed here is in accordance with these criteria. We stress again that our results also show that it is mandatory to go beyond the leading WT approximation in the interaction kernel and to properly include higher partial waves in the BSE. Moreover, contributions from three-particle intermediate states, such as $\pi\pi N$, might play a role in kaon photoproduction, see e.g. [136], but are beyond the scope of the present investigation and have thus been discarded. For a method to incorporate such three-particle cuts, see ref. [137].

5.5 Conclusions and outlook

The search for missing resonances is of great importance and actively being pursued at various experimental facilities. A promising tool to discover new resonances is the photoproduction of kaons on protons. A strong coupling to channels with open strangeness could reveal new resonances yet undetected in previous experiments based on pion-nucleon physics. Experiments with open strangeness are currently performed at ELSA, JLab and at SPring-8 providing a host of experimental data. The obtained data must be met by theoretical approaches and yield a set of tight constraints. In this work, a chiral unitary approach based on the combination of the chiral effective Lagrangian with a coupled-channels Bethe-Salpeter equation is presented. The method is exactly unitary and satisfies gauge invariance. It improves previous approaches in this field which employed only a subset of the Feynman diagrams considered here.

We have fitted both the differential cross sections for photoproduction into $K^+\Lambda$, $K^+\Sigma^0$, $K^0\Sigma^+$ as well as of the meson-baryon scattering processes $\pi^- p \to K^0\Lambda$, $K^0\Sigma^0$, $K^+\Sigma^-$. In the fits, we have restricted ourselves to the threshold regions of the respective channels, as we cannot expect

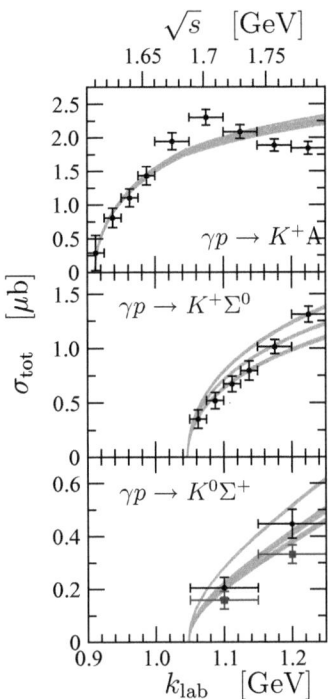

Figure 5.11: Total cross sections for photoproduction of $K^+\Lambda$ (top), $K^+\Sigma^0$ (middle), $K^0\Sigma^+$ (bottom) as a function of the photon lab momentum k_{lab}. The corresponding c.m. energy $\sqrt{s}$ is also given. The data are taken from [110] (circles) and [113] (squares).

to obtain a realistic description of these processes away from threshold due to the increasing importance of higher partial waves absent in the Weinberg-Tomozawa interaction kernel. In order to improve the agreement with experimental data, the inclusion of higher chiral orders in the interaction kernel is necessary. In charged kaon photoproduction, important contributions to the p and higher partial waves are generated by the kaon pole term, which is already included at the order we are working.

Our aim was to estimate, on the one hand, the size of gauge invariance violations introduced in previous coupled-channels analyses by neglecting Feynman diagrams where the photon couples to intermediate states of the Bethe-Salpeter bubble chain. On the other hand, the simplification of setting the interaction kernel on-shell and performing s-wave projections on the meson-baryon and the photon-baryon subsystems is studied. Our investigation suggests that both approximations are not justified in the treatment of photoproduction processes and lead to sizable changes in the results. It is thus important to satisfy gauge invariance and include off-shell terms in the effective potential. The approach presented here fulfills these requirements in a minimal way, i.e. any subset of the included diagrams cannot be omitted as this would

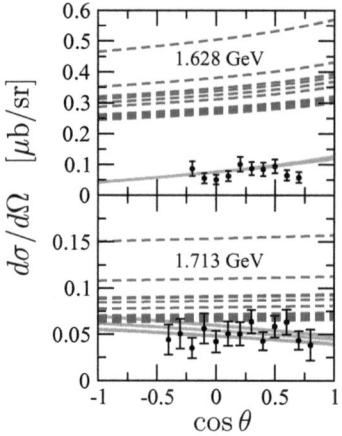 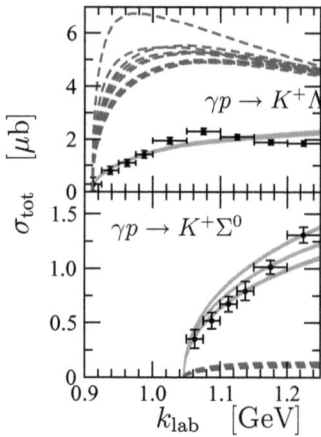

Figure 5.12: Sample results which illustrate the importance of contributions which are needed to fulfill gauge invariance. Left panel: differential cross sections for photoproduction of $K^+\Lambda$ (top), $K^+\Sigma^0$ (bottom) at c.m. energies stated in the figures compared to data from [112]. Right panel: total cross sections for photoproduction of $K^+\Lambda$ (top), $K^+\Sigma^0$ (bottom) compared to data from [109]. The full result is represented by the solid lines, whereas the dashed lines indicate the contributions from diagrams A+F in fig. 5.3.

violate either exact unitarity or gauge invariance.

There are three directions in which the approach presented here needs to be improved: First, the inclusion of higher order terms in the interaction kernel and the resulting better inclusion of p-waves and higher multipoles is straightforward but tedious. This will allow to include also the polarization data into the analysis. Note that such higher order terms, in particular the dimension two magnetic couplings $\sim \sigma_{\mu\nu}F^{\mu\nu}$, are known to be important from earlier phenomenological studies of meson electroproduction. In this context, we should also mention that in the chiral unitary approach not all resonances are generated dynamically, so in fact one might have to include explicit resonance fields in certain channels to achieve a precise description. Such a method has already been developed for elastic pion-nucleon scattering, see ref. [57]. Second, one should also include pion and photon induced η, η' production data as already accomplished in [91] and further constrain the scattering and production amplitudes through matching to the two-flavor sector (as done e.g. in refs. [138, 139]). Last but not least, the violation of crossing symmetry in the usual BSE-framework (see e.g. sec. 1.5) needs to be repaired. This could, in principle, be done by formulating Roy-type equations, but is technically involved.

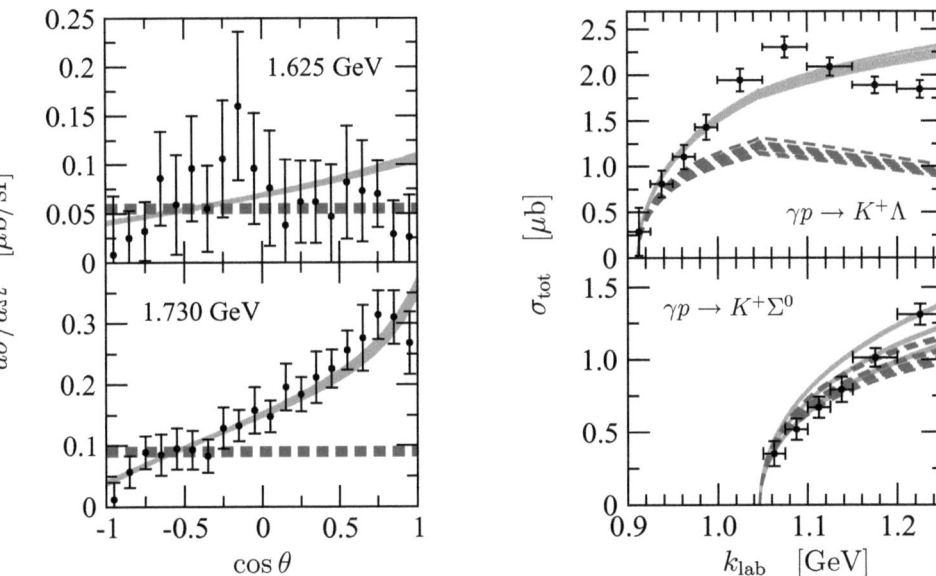

Figure 5.13: Sample results which illustrate the importance of contributions neglected in conventional coupled-channels analyses. Left panel: differential cross sections for photoproduction of $K^+\Lambda$ at two different c.m. energies as stated in the figures. Right panel: total cross sections for photoproduction of $K^+\Lambda$ (top), $K^+\Sigma^0$ (bottom). The full result is represented by the solid lines, whereas the dashed lines correspond to the outcome of the model [91] given the same values of the fit parameters. The data are taken from [109].

Chapter 6

Outlook

In this chapter, we shall take the opportunity to indicate some directions in which the work presented in this thesis can be extended. It is evident that the contributions collected here can only provide some first steps towards a systematic extension of the range of applicability of ChPT. The limits of applicability of the low-energy effective field theory are usually determined by resonances. For example, the presence of the lowest-lying vector meson octet sets the energy scale for the breakdown of ChPT in the meson sector: p-wave Goldstone-boson scattering at such energies can certainly not be correctly described by a perturbative treatment employing the effective Goldstone-boson Lagrangian. There are basically two ways in which resonances can enter the theoretical description of the effective theory: In chapters 2 and 3, we dealt with chiral Lagrangians with explicitly included resonance fields, whereas in chapters 4 and 5, we were concerned with so-called chiral unitary approaches, in which some resonances can be generated dynamically from the underlying meson-meson or meson-baryon interaction. Both those ways of implementing resonances in the description are nothing but resummation schemes, where some physical principles are used to guess the form of higher-order terms that one cannot rigorously compute in the strict perturbative approach to the effective theory. In view of the corresponding lack of theoretical rigour, the success of such resummation schemes in the description of the phenomenology is quite impressive in many cases. Let us, however, make some remarks concerning possible improvements of the theoretical work presented in the foregoing chapters.

(1) Infrared regularization with resonances: First of all, the variants of the standard infrared regularization scheme developed in chapters 2 and 3 can be applied to many other interesting observables where resonances are supposed to play an important role. Some examples have been mentioned in sec. 2.1, and at the end of chapter 3. Concerning the analysis in the latter chapter, it would be interesting to repeat that calculation with the Δ-field included explicitly. However, this is presumably complicated due to the fact that there is no clear separation of scales in that situation. For example, it is not clear a priori whether the mass splitting between the Roper and the Δ should be considered as a small or a large quantity in the counting scheme (see the discussion at the end of sec. 3.3). We should also stress that the infrared regularization scheme involving resonances is presently limited to one-loop amplitudes. Though it may be only of academic interest, it would nonetheless be reassuring to see that the method can in principle be extended to multi-loop (or at least two-loop) amplitudes.

(2) Gauge invariant chiral unitary approaches: The analysis of kaon photoproduction in chapter 5, based on the general method outlined in chapter 4, only made use of the minimal form of an amplitude subject to the constraints of two-body (coupled channel) unitarity and gauge-invariance. As expected, we could not produce a perfect fit to all the available experimental data, but nonetheless the result was encouraging, and it seems worthwhile to work on improvements of the minimal ansatz for the amplitude. The most important building block is the amplitude for meson-baryon scattering, given by the solution of a BSE with a given kernel. So far, this kernel was given only by the Weinberg-Tomozawa term, and it is not surprising that our description of meson-baryon scattering is not yet satisfactory in most channels. It is straightforward (though still requiring quite some amount of work) to implement higher-order contact terms from the next-to-leading order meson-baryon Lagrangian in the kernel (of course, this will also lead to additional photon-couplings by the minimal substitution rule). This should be the first step in future improvements. However, at this point it seems to be inconsistent to neglect the Born terms in the meson-baryon potential. Unfortunately, the iteration of the u-channel Born term leads to multiloop topologies (and three-body intermediate states in the resummed graphs) and does not seem to be feasible without making some approximation. Moreover, the coupling of the photon in loop graphs with such u-channel insertions will lead to even more complicated graphs. Therefore, one has to solve the problem of inventing some sensible approximation for such terms, which must also be consistent with unitarity and gauge invariance. This is clearly a formidable task, which must inevitably be solved to extend the present work in a systematic way. Having achieved a suitable extension of the meson-baryon scattering amplitude, the photoproduction amplitude can be worked out along the lines displayed pictorially in figs. 5.2 and 5.3. Furthermore, the coupling of the photon to the mesons and baryons will have to be supplemented with higher-order terms, like e.g. couplings proportional to $\sigma^{\mu\nu}F_{\mu\nu}$. Finally, the full restoration of crossing symmetry is obviously impossible in the BSE-framework, and requires different techniques like Roy-type equations. It is probably fair to say that the application of such methods to photoproduction processes will not be accomplished in the very near future. On the other hand, we believe that refined versions of the 'turtle' model of chapter 5 will produce sensible results in the near-threshold regions, and will yield an important contribution to the theoretical understanding of the very complicated process of low-energy kaon photoproduction.

Appendix A

Explicit expression for I_{MV}^{IR}

In eq. (8.4) of ref. [51], a closed expression for I_{MV}^{IR} for $d \to 4$ was given that looks much simpler than our result, eq. (2.15), where there is still an infinite sum to be performed. On the other hand, it is quite tedious to work out the chiral expansion of the result of [51], due to the rather complicated expressions

$$x_{1,2} = \frac{1}{2\tilde{\beta}}\left(\tilde{\beta} + \tilde{\alpha} - 1 \pm \sqrt{(\tilde{\beta} + \tilde{\alpha} - 1)^2 - 4\tilde{\alpha}\tilde{\beta}}\right) \tag{A.1}$$

used there. These two expressions are nothing but the zeroes of the Feynman parameter integral encountered in the computation of the loop integral. The corresponding chiral expansions start with

$$x_1 = -\tilde{\alpha} + \dots,$$
$$x_2^{-1} = -\tilde{\beta} + \dots,$$

showing that x_1 is small and negative, while x_2 tends to minus infinity for $\tilde{\beta} \to 0_+$. In order to show the equivalence of the two results for I_{MV}^{IR}, we employ the following relations,

$$\tilde{\beta}(x_1 + x_2) = \tilde{\beta} + \tilde{\alpha} - 1, \tag{A.2}$$
$$\tilde{\beta} x_1 x_2 = \tilde{\alpha}, \tag{A.3}$$

to write

$$\frac{(\tilde{\alpha}\tilde{\beta})^k}{(1 - \tilde{\alpha} - \tilde{\beta})^{2k+1}} = -\frac{1}{\tilde{\beta} x_2} \frac{(\frac{x_1}{x_2})^k}{(1 + \frac{x_1}{x_2})^{2k+1}} = -\frac{1}{\tilde{\beta} x_2} \sum_{m=0}^{\infty} \frac{(-1)^m \Gamma(2k+m+1)}{m! \Gamma(2k+1)} \left(\frac{x_1}{x_2}\right)^{k+m}.$$

Inserting this in eq. (2.15) yields

$$\begin{aligned}
I_{MV}^{IR}(q^2) &= \frac{M_V^{d-4}(\tilde{\alpha})^{\frac{d}{2}-1}}{(4\pi)^{\frac{d}{2}}\tilde{\beta} x_2} \sum_{k=0}^{\infty}\sum_{m=0}^{\infty} \frac{\Gamma(2k+m+1)\Gamma(1-\frac{d}{2}-k)\Gamma(2k+1)}{\Gamma(2k+1)\Gamma(k+1)\Gamma(m+1)} \left(-\frac{x_1}{x_2}\right)^{k+m} \\
&= \frac{M_V^{d-4}(\tilde{\alpha})^{\frac{d}{2}-2} x_1}{(4\pi)^{\frac{d}{2}}} \sum_{k=0}^{\infty}\sum_{m=0}^{\infty} \frac{\Gamma(2k+m+1)\Gamma(1-\frac{d}{2}-k)}{\Gamma(k+1)\Gamma(m+1)} \left(-\frac{x_1}{x_2}\right)^{k+m}.
\end{aligned} \tag{A.4}$$

In the second line we made use of eq. (A.3). Now we change the summation indices according to
$$j = k+m, \qquad l = k = j-m,$$
and use the following sum formula for Gamma functions,
$$\sum_{l=0}^{j} \frac{\Gamma(j+l+1)\Gamma(x-l)}{\Gamma(j-l+1)\Gamma(l+1)} = (-1)^j \frac{\Gamma(x-j)\Gamma(-x)}{\Gamma(-x-j)}, \qquad j \in \mathbb{N}, \qquad (A.5)$$
for $x = 1 - \frac{d}{2}$. We shall give a short outline of a proof for eq. (A.5): Dividing this equation by $\Gamma(x-j)$, both sides are just polynomials in x of degree j, with coefficient 1 in front of x^j. Consequently, one only has to prove that both polynomials have the same set of zeroes, namely $\{-1, -2, \ldots, -j\}$. This is not difficult, making use of
$$\sum_{l=i}^{j} \frac{(-1)^{j-l}\Gamma(j-i+1)}{\Gamma(j-l+1)\Gamma(l+1)} \prod_{p=0}^{i-1}(l-p) = \sum_{l=i}^{j} \frac{(-1)^{j-l}\Gamma(j-i+1)}{\Gamma(j-l+1)\Gamma(l-i+1)}$$
$$= \sum_{n=0}^{j-i} \binom{j-i}{n}(-1)^{j-i-n} = (1-1)^{j-i} = 0$$
for $0 < i < j$. These remarks should be sufficient to complete the proof of eq. (A.5). Returning to eq. (A.4), we employ eq. (A.5) to write
$$I_{MV}^{IR}(q^2) = \frac{M_V^{d-4}(\tilde{\alpha})^{\frac{d}{2}-2}x_1}{(4\pi)^{\frac{d}{2}}} \sum_{j=0}^{\infty} \sum_{l=0}^{j} \frac{\Gamma(j+l+1)\Gamma(1-\frac{d}{2}-l)}{\Gamma(j-l+1)\Gamma(l+1)} \left(-\frac{x_1}{x_2}\right)^j$$
$$= \frac{M_V^{d-4}(\tilde{\alpha})^{\frac{d}{2}-2}x_1}{(4\pi)^{\frac{d}{2}}} \sum_{j=0}^{\infty} \frac{\Gamma(\frac{d}{2}-1)\Gamma(1-\frac{d}{2}-j)}{\Gamma(\frac{d}{2}-1-j)}\left(\frac{x_1}{x_2}\right)^j. \qquad (A.6)$$

The last line of eq. (A.6) is exactly the result that was derived in sections 7 and 8 of [51]. This can be seen by substituting
$$a \to -x_1, \quad b \to -x_2^{-1}, \quad d \to \frac{d}{2} - 2$$
in eq. (7.7) of that reference, and multiplying the result with
$$-\frac{\Gamma(2-\frac{d}{2})M_V^{d-4}}{(4\pi)^{\frac{d}{2}}}(-\tilde{\beta}x_2)^{\frac{d}{2}-2},$$
as explained at the beginning of sec. 8 of [51]. In the limit $d \to 4$, the series of eq. (A.6) can be summed up to give
$$I_{MV}^{IR}(q^2, d \to 4) = 2x_1\lambda - \frac{1}{16\pi^2}\left(x_1(1-\ln\tilde{\alpha}) - (x_1-x_2)\ln\left(1-\frac{x_1}{x_2}\right)\right), \qquad (A.7)$$
where
$$\lambda = \frac{M_V^{d-4}}{16\pi^2}\left(\frac{1}{d-4} - \frac{1}{2}(\ln(4\pi) - \gamma + 1)\right).$$
Eq. (A.7) is identical to eq. (8.4) of [51].

Appendix B

Alternative derivation of I_{MBV}^{IR}

In this appendix, we present an alternative derivation of the infrared singular part of the loop integral $I_{MBV}(k^2)$ (see eqs. (2.20,2.27)) using the prescription of Ellis and Tang we have already explained at the end of sec. 2.3. However, at some places we will also use Feynman parameter integrals, so the derivation outlined here is to some extent a mixture of the standard infrared regularization procedure and the method of Ellis and Tang. In complete analogy to the steps performed at the end of sec. 2.3, we start with

$$
\begin{aligned}
I_{MBV}(k^2) &\to \int \frac{d^d l}{(2\pi)^d} \frac{i}{((p-l)^2 - m^2)(l^2 - M^2)} \sum_{j=0}^{\infty} \frac{(-2k \cdot l)^j}{(k^2 + l^2 - M_V^2)^{j+1}} \\
&\to \int \frac{d^d l}{(2\pi)^d} \frac{i}{((p-l)^2 - m^2)(l^2 - M^2)} \sum_{j=0}^{\infty} \frac{(-2k \cdot l)^j}{(k^2 + M^2 - M_V^2)^{j+1}} \\
&\to \sum_{j=0}^{\infty} \int \frac{d^d l}{(2\pi)^d} \frac{i(-2k \cdot l)^j}{(k^2 + M^2 - M_V^2)^{j+1}((p-l)^2 - m^2)(l^2 - M^2)}.
\end{aligned}
\tag{B.1}
$$

Now we could use the procedure outlined in [39] to expand the nucleon propagator, together with an interchange of summation and integration. In the present case, however, it is easy to see that it is equivalent to use the common Feynman parameter trick for the remaining loop integrals and extend the parameter integration to infinity like in sec. 2.2. For I_{MB} (see eqs. (2.1) and (2.3)), the splitting of eq. (2.5) corresponds to

$$
\frac{1}{((p-l)^2 - m^2)(l^2 - M^2)} = \frac{1}{p^2 - m^2 - 2l \cdot p + M^2} \left(\frac{1}{l^2 - M^2} - \frac{1}{(p-l)^2 - m^2} \right). \tag{B.2}
$$

This has also been noted in [38], see eqs. (22,23) of that reference. On the other hand, the first term on the r.h.s of eq. (B.2) is exactly the integrand that gives the soft momentum contribution in the sense of Ellis and Tang, see e.g. eq. (7) in [39]. Therefore, it is consistent to continue the series of steps in eq. (B.1) with

$$
\begin{aligned}
\ldots &\to \sum_{j=0}^{\infty} \int_0^{\infty} \frac{dz}{(k^2 + M^2 - M_V^2)^{j+1}} \int \frac{d^d l}{(2\pi)^d} \frac{i(-2k \cdot l)^j}{[((p-l)^2 - m^2)z + (l^2 - M^2)(1-z)]^2} \\
&\equiv I_{MBV}^{\text{soft}}(k^2).
\end{aligned}
\tag{B.3}
$$

The loop-integration can be done using the following generalization of eq. (2.16):

$$\int \frac{d^d l}{(2\pi)^d} \frac{i(k \cdot l)^{2n}}{(l^2 - M^2)^r} = (-1)^{r-1}(-k^2 M^2)^n M^{d-2r} \frac{\Gamma(n+\frac{1}{2})}{\Gamma(\frac{1}{2})} \frac{\Gamma(r - \frac{d}{2} - n)}{(4\pi)^{\frac{d}{2}} \Gamma(r)} \qquad (B.4)$$

for $r, n \in \mathbb{N}$. This gives

$$I^{\text{soft}}_{MVB}(k^2) = \sum_{j=0}^{\infty} \sum_{i=0, i \in 2\mathbb{N}}^{j} \frac{m^{d-6} 2^i \binom{j}{i}(-\beta)^{j-\frac{i}{2}}}{(\gamma - \alpha - \beta)^{j+1}} \frac{\Gamma(\frac{i+1}{2})\Gamma(2 - \frac{d}{2} - \frac{i}{2})}{\Gamma(\frac{1}{2})(4\pi)^{\frac{d}{2}}} \int_0^{\infty} \frac{dz\, z^{j-i}}{(z^2 - \alpha z + \alpha)^{2 - \frac{d}{2} - \frac{i}{2}}}.$$

Here we have also used the on-shell kinematics specified in eq. (2.19). The sum over the even integers i extends only to $j-1$ if j is odd. Defining new indices,

$$J = j - \frac{i}{2}, \qquad l = \frac{i}{2},$$

and reordering the series correspondingly gives the following expression for $I^{\text{soft}}_{MVB}(k^2)$:

$$\sum_{J=0}^{\infty} \sum_{l=0}^{J} \frac{m^{d-6}(-\beta)^J 4^l \Gamma(J+l+1)\Gamma(l+\frac{1}{2})\Gamma(2-\frac{d}{2}-l)}{(4\pi)^{\frac{d}{2}}(\gamma-\alpha-\beta)^{J+l+1}\Gamma(J-l+1)\Gamma(2l+1)\Gamma(\frac{1}{2})} \int_0^{\infty} \frac{z^{J-l} dz}{(z^2 - \alpha z + \alpha)^{2-\frac{d}{2}-l}}.$$

Using eq. (2.18), it is straightforward to see that this equals I^{IR}_{MBV} of eq. (2.27), as expected (the remaining parameter integral can be done with the help of eq. (2.26)).

Appendix C

Decomposition of infrared regularized loop integrals

Here we list the decomposition of the loop integrals with tensor structures in the numerator, which we need in sec. 2.5. All loop integrals in this appendix are understood as the infrared singular parts of the full loop integrals, but we will suppress the superscript IR for brevity. As a consequence, all loop integrals that do not contain a pion propagator are already dropped here, since they have no infrared singular part. Also, we will use the mass shell condition $p^2 = m^2$ for the nucleon momentum p. We start with

$$\int \frac{d^d l}{(2\pi)^d} \frac{il^\mu}{((p-l)^2 - m^2)(l^2 - M^2)} = p^\mu I_{MB}^{(1)}, \tag{C.1}$$

where

$$I_{MB}^{(1)} = \frac{1}{2m^2}\left(M^2 I_{MB} - I_M\right).$$

In complete analogy,

$$\int \frac{d^d l}{(2\pi)^d} \frac{il^\mu}{((l-k)^2 - M_V^2)(l^2 - M^2)} = k^\mu I_{MV}^{(1)}, \tag{C.2}$$

with

$$I_{MV}^{(1)} = \frac{1}{2k^2}\left((k^2 + M^2 - M_V^2) I_{MV} - I_M\right).$$

Integrals of type MB and MV are also needed with a tensor structure $l^\mu l^\nu$ in the numerator. They are decomposed as

$$\int \frac{d^d l}{(2\pi)^d} \frac{il^\mu l^\nu}{((l-k)^2 - M_V^2)(l^2 - M^2)} = g^{\mu\nu} t_{MV}^{(0)}(k) + \frac{k^\mu k^\nu}{k^2} t_{MV}^{(1)}(k), \tag{C.3}$$

where the coefficients of the tensor structures are given by

$$(d-1)t_{MV}^{(0)}(k) = \frac{4k^2 M^2 - (k^2 + M^2 - M_V^2)^2}{4k^2} I_{MV} + \frac{k^2 + M^2 - M_V^2}{4k^2} I_M,$$

$$(d-1)t_{MV}^{(1)}(k) = \frac{d(k^2 + M^2 - M_V^2)^2 - 4k^2 M^2}{4k^2} I_{MV} - \frac{d(k^2 + M^2 - M_V^2)}{4k^2} I_M.$$

The corresponding coefficients in the meson-baryon case, $t_{MB}^{(0,1)}(p)$, can be derived from these results by substituting $k \to p$, $M_V \to m$.

We turn now to loop integrals with three propagators. First the vector integral:

$$\int \frac{d^d l}{(2\pi)^d} \frac{i l^\mu}{((\bar{p}-l)^2 - m^2)((l-k)^2 - M_V^2)(l^2 - M^2)} = (k+\bar{p})^\mu I_{MBV}^A + (k-\bar{p})^\mu I_{MBV}^B, \quad (C.4)$$

with

$$I_{MBV}^A = \frac{1}{2k^2(4m^2 - k^2)}\Big[(2m^2 M^2 + (2m^2 - k^2)(k^2 - M_V^2))I_{MBV}$$
$$- k^2 I_{MV} - (2m^2 - k^2)I_{MB}\Big],$$

$$I_{MBV}^B = \frac{1}{2k^2(4m^2 - k^2)}\Big[(2M^2(m^2 - k^2) + (k^2 - M_V^2)(k^2 + 2m^2))I_{MBV}$$
$$+ 3k^2 I_{MV} - (k^2 + 2m^2)I_{MB}\Big].$$

We remind the reader that we use $\bar{p}^2 = m^2 = (\bar{p}-k)^2$ here. The scalar loop integral with three propagators occuring in this decomposition was named $\tilde{I}_{MBV}$ in sec. 2.4, eq. (2.23). However, we noted there that it is equal to I_{MBV} for on-shell nucleon momenta.

The tensor integral

$$I_{MBV}^{\mu\nu} = \int \frac{d^d l}{(2\pi)^d} \frac{i l^\mu l^\nu}{((\bar{p}-l)^2 - m^2)((l-k)^2 - M_V^2)(l^2 - M^2)} \quad (C.5)$$

can be decomposed as

$$I_{MBV}^{\mu\nu} = g^{\mu\nu} C_1 + (\bar{p}+k)^\mu(\bar{p}+k)^\nu C_2 + (\bar{p}-k)^\mu(\bar{p}-k)^\nu C_3 + ((\bar{p}+k)^\mu(k-\bar{p})^\nu + (\bar{p}+k)^\nu(k-\bar{p})^\mu) C_4.$$

Here the coefficients C_i are given by

$$C_1 = \frac{1}{d-2}\Big[M^2 I_{MBV} - \frac{1}{2}(k^2 - M_V^2 + 2M^2)I_{MBV}^A + \frac{M_V^2 - k^2}{2} I_{MBV}^B\Big],$$

$$C_2 = \frac{1}{k^2(4m^2 - k^2)}\Big[m^2 M^2 I_{MBV} + \frac{M_V^2 - k^2}{2}\left(m^2 I_{MBV}^B + (k^2 - m^2)I_{MBV}^A\right)$$
$$- \frac{2m^2 - k^2}{4} I_{MB}^{(1)} - \frac{k^2}{4} I_{MV}^{(1)} - m^2(d-1)C_1\Big],$$

$$C_3 = \frac{1}{k^2(4m^2 - k^2)}\Big[M^2(m^2 + 2k^2)I_{MBV} - \frac{1}{2}(k^2 - M_V^2 + 2M^2)((m^2 + 2k^2)I_{MBV}^A$$
$$+ (k^2 - m^2)I_{MBV}^B) + \frac{3k^2}{4} I_{MV}^{(1)} + \frac{2m^2 + k^2}{4} I_{MB}^{(1)} - (m^2 + 2k^2)(d-1)C_1\Big],$$

$$C_4 = \frac{1}{k^2(4m^2 - k^2)}\Big[M^2(m^2 - k^2)I_{MBV} + \frac{k^2 - M_V^2}{2}\left((2k^2 + m^2)I_{MBV}^A\right.$$
$$\left. + (k^2 - m^2)I_{MBV}^B\right) + \frac{3k^2}{4} I_{MV}^{(1)} - \frac{2m^2 + k^2}{4} I_{MB}^{(1)}$$
$$- (m^2 - k^2)(d-1)C_1\Big].$$

Appendix D
Ward identity for general interaction kernels

In this appendix we derive eq. (4.36), which follows from a contact interaction of the type

$$\mathcal{L}_{\text{int}} = \bar{B}\, C_{\mu_1\cdots\mu_l\mu_{l+1}\cdots\mu_m\mu_{m+1}\cdots\mu_n} \left(D^{\mu_1}\ldots D^{\mu_l}\tilde{\phi}\right)\left(D^{\mu_{l+1}}\ldots D^{\mu_m}\phi^\dagger\right)\left(D^{\mu_{m+1}}\ldots D^{\mu_n}\tilde{B}\right) \quad \text{(D.1)}$$

with $D_\mu = \partial_\mu + i\hat{e}A_\mu$. The $\gamma(k)\tilde{\phi}(q_i)\tilde{B}(p_i) \to \phi(q_f)B(p_f)$ vertex is obtained from this Lagrangian by extracting the part linear in $\mathcal{A}_\mu$. A partial derivative acting, e.g., on an incoming meson $\phi(q)$ yields a factor $-iq$ in momentum space, while a partial derivative on $\phi(q)\mathcal{A}(k)$ (both momenta incoming) leads to $-i(q+k)$. Therefore, one obtains the vertex

$$\begin{aligned}
A_\mu = {} & i\sum_{s=1}^{l} C_{\mu_1\cdots\mu_n}\Big|_{\mu_s=\mu} (-i)^{(2l-2m+n)} \left[\prod_{t=1}^{s-1}(q_i+k)^{\mu_t}\right](-\hat{e}_{\tilde{\phi}})\left[\prod_{t=s+1}^{l} q_i^{\mu_t}\right]\left[\prod_{t=l+1}^{m} q_f^{\mu_t}\right]\left[\prod_{t=m+1}^{n} p_i^{\mu_t}\right] \\
& + i\sum_{s=l+1}^{m} C_{\mu_1\cdots\mu_n}\Big|_{\mu_s=\mu} (-i)^{(2l-2m+n)} \left[\prod_{t=1}^{l} q_i^{\mu_t}\right]\left[\prod_{t=l+1}^{s-1}(q_f-k)^{\mu_t}\right](-\hat{e}_\phi)\left[\prod_{t=s+1}^{m} q_f^{\mu_t}\right]\left[\prod_{t=m+1}^{n} p_i^{\mu_t}\right] \\
& + i\sum_{s=m+1}^{n} C_{\mu_1\cdots\mu_n}\Big|_{\mu_s=\mu} (-i)^{(2l-2m+n)} \left[\prod_{t=1}^{l} q_i^{\mu_t}\right]\left[\prod_{t=l+1}^{m} q_f^{\mu_t}\right]\left[\prod_{t=m+1}^{s-1}(p_i+k)^{\mu_t}\right](-\hat{e}_{\tilde{B}})\left[\prod_{t=s+1}^{n} p_i^{\mu_t}\right]
\end{aligned} \quad \text{(D.2)}$$

Multiplying this equation with k_μ and making use of charge conservation at the vertex which follows from gauge invariance of the Lagrangian one arrives at

$$\begin{aligned}
k^\mu A_\mu = {} & iC_{\mu_1\cdots\mu_n}(-i)^{(2l-2m+n)}(-\hat{e}_{\tilde{\phi}})(q_i+k)^{\mu_1}\cdots(q_i+k)^{\mu_l} q_f^{\mu_{l+1}}\cdots q_f^{\mu_m} p_i^{\mu_{m+1}}\cdots p_i^{\mu_n} \\
& - iC_{\mu_1\cdots\mu_n}(-i)^{(2l-2m+n)}(-\hat{e}_\phi) q_i^{\mu_1}\cdots q_i^{\mu_l}(q_f-k)^{\mu_{l+1}}\cdots(q_f-k)^{\mu_m} p_i^{\mu_{m+1}}\cdots p_i^{\mu_n} \\
& + iC_{\mu_1\cdots\mu_n}(-i)^{(2l-2m+n)}(-\hat{e}_{\tilde{B}}) q_i^{\mu_1}\cdots q_i^{\mu_l} q_f^{\mu_{l+1}}\cdots q_f^{\mu_m}(p_i+k)^{\mu_{m+1}}\cdots(p_i+k)^{\mu_n} \\
& - iC_{\mu_1\cdots\mu_n}(-i)^{(2l-2m+n)}(-\hat{e}_B) q_i^{\mu_1}\cdots q_i^{\mu_l} q_f^{\mu_{l+1}}\cdots q_f^{\mu_m} p_i^{\mu_{m+1}}\cdots p_i^{\mu_n}\, .
\end{aligned} \quad \text{(D.3)}$$

On the other hand, the vertex $\tilde{\phi}(q_i)\tilde{B}(p_i) \to \phi(q_f)B(p_f)$ from the Lagrangian (D.1) is given by the piece without the photon field $\mathcal{A}_\mu$ and reads

$$A(p_i, q_i, q_f) = iC_{\mu_1\cdots\mu_n}(-i)^{(2l-2m+n)} q_i^{\mu_1}\cdots q_i^{\mu_l} q_f^{\mu_{l+1}}\cdots q_f^{\mu_m} p_i^{\mu_{m+1}}\cdots p_i^{\mu_n}\, . \quad \text{(D.4)}$$

This proves the Ward identity (4.36)

$$\begin{aligned}
k^\mu A_\mu(p_i, q_i, q_f, k) = {} & \hat{e}_\phi A(p_i, q_i, q_f - k) - \hat{e}_{\tilde{\phi}} A(p_i, q_i + k, q_f) \\
& + \hat{e}_B A(p_i, q_i, q_f) - \hat{e}_{\tilde{B}} A(p_i + k, q_i, q_f)\, .
\end{aligned} \quad \text{(D.5)}$$

Appendix E
Solution of the Bethe-Salpeter equation

The solution of the BSE for the Weinberg-Tomozawa potential

$$V(\slashed{q}_2, \slashed{q}_1) = g(\slashed{q}_1 + \slashed{q}_2), \tag{E.1}$$

$$T(\slashed{q}_2, \slashed{q}_1; p) = V(\slashed{q}_2, \slashed{q}_1) + \int \frac{d^d l}{(2\pi)^d} V(\slashed{q}_2, \slashed{l}) i S(\slashed{p} - \slashed{l}) \Delta(l) T(\slashed{l}, \slashed{q}_1; p), \tag{E.2}$$

was claimed in eq. (5.6) to be

$$T(\slashed{q}_2, \slashed{q}_1; p) = W(\slashed{q}_2, \slashed{q}_1; p) + W(\slashed{q}_2, \tilde{p}; p) G(p)[1 - W(\tilde{p}, \tilde{p}; p) G(p)]^{-1} W(\tilde{p}, \slashed{q}_1; p), \tag{E.3}$$

with

$$W(\slashed{q}_2, \slashed{q}_1; p) = \slashed{q}_2 g \frac{1}{1 + I_M g} + \frac{1}{1 + g I_M} g \slashed{q}_1 - g \frac{1}{1 + I_M g} I_M (\slashed{p} - m) \frac{1}{1 + g I_M} g. \tag{E.4}$$

$W(\slashed{q}_2, \tilde{p}; p)$ can be obtained from eq. (E.4) by replacing $\slashed{q}_1 \to \slashed{q}_1 + \slashed{p}_1 - m = \slashed{p} - m \equiv \tilde{\slashed{p}}$, and similarly for other arguments. Recall that m is the baryon mass matrix, and p_1 (p_2) is the four-momentum of the incoming (outgoing) baryon. The propagators S and Δ, as well as the loop integrals I_M and G, have been defined in eqs. (5.3,5.4) and eqs. (5.9,5.10), respectively. Note that I_M and G both commute with m, since these are all diagonal matrices in channel space. In contrast, the coupling matrix g does not commute with any of those matrices. We also remind the reader that the symbol '1' in eqs. (E.3) and (E.4) represents the unit matrix in channel space, while matrix-valued denominators denote matrix inversion. In order to verify that the above expression for T indeed solves the BSE, we need to prove the relation

$$\int \frac{d^d l}{(2\pi)^d} V(\slashed{q}_2, \slashed{l}) i S(\slashed{p} - \slashed{l}) \Delta(l) W(\slashed{l}, \slashed{q}; p)$$
$$= V(\slashed{q}_2, \tilde{p}) G(p) W(\tilde{p}, \slashed{q}; p) + W(\slashed{q}_2, \slashed{q}; p) - V(\slashed{q}_2, \slashed{q}). \tag{E.5}$$

where $\slashed{q}$ stands for either $\slashed{q}_1$ or $\tilde{p}$. To proceed, we split the loop momentum l in two pieces according to $\slashed{l} = \tilde{\slashed{p}} - (\tilde{\slashed{p}} - \slashed{l})$. The second part in this decomposition cancels the denominator of the baryon propagator $S(\slashed{p} - \slashed{l})$. One finds

$$\int \frac{d^d l}{(2\pi)^d} V(\slashed{q}_2, \slashed{l}) i S(\slashed{p} - \slashed{l}) \Delta(l) W(\slashed{l}, \slashed{q}; p) = V(\slashed{q}_2, \tilde{p}) G(p) W(\tilde{p}, \slashed{q}; p)$$
$$- V(\slashed{q}_2, \tilde{p}) I_M g \frac{1}{1 + I_M g} - g I_M W(\tilde{p}, \slashed{q}; p) + g \tilde{\slashed{p}} I_M g \frac{1}{1 + I_M g}.$$

Using the identities

$$I_M g \frac{1}{1+I_M g} = 1 - \frac{1}{1+I_M g}, \qquad g I_M \frac{1}{1+g I_M} = 1 - \frac{1}{1+g I_M}$$

and $g[1+I_M g]^{-1} = [1+g I_M]^{-1} g$ in the last three terms then yields

$$V(\not{q}_2,\tilde{p}) I_M g \frac{1}{1+I_M g} + g I_M W(\tilde{p},\not{q};p) - g\tilde{p} I_M g \frac{1}{1+I_M g} = V(\not{q}_2,\not{q}) - W(\not{q}_2,\not{q};p),$$

which proves eq. (E.5). Inserting the expression for T, eq. (E.3), on the r.h.s. of the BSE, eq. (E.2), we get

$$\int \frac{d^d l}{(2\pi)^d} V(\not{q}_2,\not{l}) i S(\not{p}-\not{l}) \Delta(l) T(\not{l},\not{q}_1;p) = \int \frac{d^d l}{(2\pi)^d} V(\not{q}_2,\not{l}) i S(\not{p}-\not{l}) \Delta(l) W(\not{l},\not{q}_1;p)$$
$$+ \int \frac{d^d l}{(2\pi)^d} V(\not{q}_2,\not{l}) i S(\not{p}-\not{l}) \Delta(l) W(\not{l},\tilde{p};p) G(p) [1-W(\tilde{p},\tilde{p};p)G(p)]^{-1} W(\tilde{p},\not{q}_1;p). \quad \text{(E.6)}$$

Now eq. (E.5) can be used in both terms on the r.h.s. of the last equation, to obtain

$$\int \frac{d^d l}{(2\pi)^d} V(\not{q}_2,\not{l}) i S(\not{p}-\not{l}) \Delta(l) T(\not{l},\not{q}_1;p) = V(\not{q}_2,\tilde{p}) G(p) W(\tilde{p},\not{q}_1;p) + W(\not{q}_2,\not{q}_1;p) - V(\not{q}_2,\not{q}_1)$$
$$+ [V(\not{q}_2,\tilde{p})G(p)W(\tilde{p},\tilde{p};p) + W(\not{q}_2,\tilde{p};p) - V(\not{q}_2,\tilde{p})] G(p) [1-W(\tilde{p},\tilde{p};p)G(p)]^{-1} W(\tilde{p},\not{q}_1;p)$$
$$= T(\not{q}_2,\not{q}_1;p) - V(\not{q}_2,\not{q}_1), \quad \text{(E.7)}$$

where in the second step we have used $WG[1-WG]^{-1} = [1-WG]^{-1} - 1$, together with eq. (E.3). This completes the proof that T, as given in eq. (E.3), solves the integral equation (E.2).

Appendix F
UV-divergences in the Bethe-Salpeter equation

We already noted in sec. 1.5 that the loop integration in the BSE develops an ultraviolet divergence as the dimension d approaches its physical value 4. Therefore, divergent expressions will appear in the solution T of the BSE. In fact, the expression for T in the previous app. E contains the divergent loop integrals G and I_M, so that it is, strictly speaking, ill-defined at $d = 4$. If g is kept fixed, one might argue from the form of T (see eqs. (E.3) and (E.4)) that $T \to 0$ as $d \to 4$, but this cannot be the result for the sum of bubble graphs, as displayed e.g. in fig. 5.1. In practice, one uses the form of the solution of the BSE, but with finite loop functions $G - \delta G, I_M - \delta I_M$, where δG and δI_M are real polynomials in the momentum variables, which contain the poles in $d - 4$ of the respective loop integrals. In the following, we will leave out the overall momentum p in the arguments of the various functions for brevity. For the same reason, we will denote $W(\tilde{p}, \tilde{p}; p)$ simply by the letter W.

Due to the above-mentioned form of the subtractions $\delta G, \delta I_M$, it is clear that the unitarity and analyticity requirements are not spoilt by the replacement $T \to T_\delta$, where T_δ contains only finite loop functions:

$$T_\delta(\not{q}_2, \not{q}_1) = \overline{W}(\not{q}_2, \not{q}_1) + \overline{W}(\not{q}_2, \tilde{p})(G - \delta G)[1 - \overline{W}(G - \delta G)]^{-1}\overline{W}(\tilde{p}, \not{q}_1), \qquad (F.1)$$

where

$$\overline{W}(\not{q}_2, \not{q}_1) = \not{q}_2 g \frac{1}{1 + (I_M - \delta I_M)g} + \frac{1}{1 + g(I_M - \delta I_M)}g\not{q}_1$$
$$- g\frac{1}{1 + (I_M - \delta I_M)g}(I_M - \delta I_M)(\not{p} - m)\frac{1}{1 + g(I_M - \delta I_M)}g. \qquad (F.2)$$

It is the purpose of this appendix to demonstrate that T_δ is again the solution of a BSE, with a modified potential $V + \delta V \equiv V_\delta$, which can be expressed through the subtractions $\delta G, \delta I_M$ as follows:

$$V_\delta(\not{q}_2, \not{q}_1) = W_\delta(\not{q}_2, \not{q}_1) - W_\delta(\not{q}_2, \tilde{p})\delta G[1 + W_\delta \delta G]^{-1}W_\delta(\tilde{p}, \not{q}_1), \qquad (F.3)$$

where

$$W_\delta(\not{q}_2, \not{q}_1) = \not{q}_2 g\frac{1}{1 - \delta I_M g} + \frac{1}{1 - g\delta I_M}g\not{q}_1 + g\frac{1}{1 - \delta I_M g}\delta I_M(\not{p} - m)\frac{1}{1 - g\delta I_M}g. \qquad (F.4)$$

Note that V_δ has the same form as T, with G and I_M replaced by $-\delta G$ and $-\delta I_M$, respectively (compare eq. (1.49) for a simpler version of this property of the BSE. There, $G(s_0)$ plays the role of δG).

The addition of the 'counterterms' δV to the original potential V is implemented graphically by iterating also the vertices corresponding to δV: for every graph with a V-vertex at a certain position, one must also include a graph with an insertion of δV at the same position. In this way, the divergences present in the loop integrals contained in the solution of the original BSE are shifted to the kernel. In our opinion, this is the closest one can come to a renormalization procedure in the framework of the BSE. Nevertheless, let us repeat the remark already made in secs. 1.5 and 5.2, saying that the terms in δV do not correspond to counterterms as employed in perturbative ChPT. At the one-loop level, for example, the difference resides in those parts of the counterterms of ChPT which absorb the divergences of graphs with loops in the $u-$ or $t-$channel, as the latter are not present in the BSE approach.

As a side remark, we also note that our on-shell-amplitude, where $\not{q}_{1,2} \to \tilde{p}$, reduces to the result one would derive using the on-shell approximation (see the discussion in sec. 5.2) for the choice $\delta I_M = I_M$. This may be seen as the justification for the use of the on-shell approximation when considering meson-baryon scattering with the kernel V. However, it is important to note that the off-shell dependence of T (or T_δ) has to be retained when the scattering amplitude provides only a building block in more complicated graphs, as it happens in our treatment of meson photoproduction in chapter 5: Knowing only the on-shell scattering amplitude, we would not be able to use the tools developed in that chapter in order to obtain a gauge-invariant description of meson photoproduction.

Let us now proceed with the proof. We must show that

$$T_\delta(\not{q}_2,\not{q}_1) = V_\delta(\not{q}_2,\not{q}_1) + \int \frac{d^d l}{(2\pi)^d} V_\delta(\not{q}_2, \not{l}) i S(\not{p} - \not{l}) \Delta(l) T_\delta(\not{l}, \not{q}_1) \,. \tag{F.5}$$

To do this, we need to employ the relation

$$\int \frac{d^d l}{(2\pi)^d} W_\delta(\not{q}_2, \not{l}) i S(\not{p} - \not{l}) \Delta(l) \overline{W}(\not{l}, \not{q}) = W_\delta(\not{q}_2, \tilde{p}) G \overline{W}(\tilde{p}, \not{q}) + \overline{W}(\not{q}_2, \not{q}) - W_\delta(\not{q}_2, \not{q}) \,, \tag{F.6}$$

where $\not{q}$ can stand for $\not{q}_1$ or $\tilde{p}$. The proof of this identity will be deferred to the end of this appendix. For the moment, let us assume we have shown that eq. (F.6) holds true. Then we can use it to calculate (see eqs. (F.1) and (F.3))

$$\begin{aligned}
\int \frac{d^d l}{(2\pi)^d} V_\delta(\not{q}_2,\not{l}) i S(\not{p}-\not{l})\Delta(l) T_\delta(\not{l},\not{q}_1) &= W_\delta(\not{q}_2,\tilde{p}) G\overline{W}(\tilde{p},\not{q}_1) + \underline{\overline{W}(\not{q}_2,\not{q}_1)} - \underline{W_\delta(\not{q}_2,\not{q}_1)} \\
&\quad - W_\delta(\not{q}_2,\tilde{p})\delta G[1+W_\delta \delta G]^{-1}\{W_\delta G\overline{W}(\tilde{p},\not{q}_1) + \overline{W}(\tilde{p},\not{q}_1) - \underline{W_\delta(\tilde{p},\not{q}_1)}\} \\
&\quad + \{W_\delta(\not{q}_2,\tilde{p})G\overline{W} + \overline{W}(\not{q}_2,\tilde{p}) - W_\delta(\not{q}_2,\tilde{p})\}(G-\delta G)[1-\overline{W}(G-\delta G)]^{-1}\overline{W}(\tilde{p},\not{q}_1) \\
&\quad - W_\delta(\not{q}_2,\tilde{p})\delta G[1+W_\delta \delta G]^{-1}\{W_\delta G\overline{W} + \overline{W} - W_\delta\}(G-\delta G)[1-\overline{W}(G-\delta G)]^{-1}\overline{W}(\tilde{p},\not{q}_1) \,.
\end{aligned}$$

$$\tag{F.7}$$

The underlined terms lead to the result $T_\delta(\not{q}_2,\not{q}_1) - V_\delta(\not{q}_2,\not{q}_1)$, while the remaining terms on the r.h.s. of eq. (F.7) add up to zero (this can be seen by straightforward calculation, employing simple algebraic identities analogous to the ones used in the proof of the previous appendix). Recall that omitted arguments of the W-functions indicate that both arguments equal $\tilde{p}$.

To complete the proof of eq. (F.5), we still have to verify eq. (F.6). Let us evaluate the integral on the l.h.s. of that equation, inserting the appropriate expressions for $\overline{W}$ and W_δ:

$$\int \frac{d^d l}{(2\pi)^d} W_\delta(\slashed{q}_2, \slashed{l}) iS(\slashed{p} - \slashed{l}) \Delta(l) \overline{W}(\slashed{l}, \slashed{q}) = W_\delta(\slashed{q}_2, \tilde{p}) G \overline{W}(\tilde{p}, \slashed{q}) - \frac{1}{1 - g\delta I_M} g I_M \overline{W}(\tilde{p}, \slashed{q})$$
$$- W_\delta(\slashed{q}_2, \tilde{p}) I_M g \frac{1}{1 + (I_M - \delta I_M)g} + \frac{1}{1 - g\delta I_M} g I_M \tilde{p} g \frac{1}{1 + (I_M - \delta I_M)g} . \quad (F.8)$$

The first term on the r.h.s. already coincides with the one in eq. (F.6). It remains to be shown that

$$\frac{1}{1 - g\delta I_M} g I_M \overline{W}(\tilde{p}, \slashed{q}) + W_\delta(\slashed{q}_2, \tilde{p}) I_M g \frac{1}{1 + (I_M - \delta I_M)g} - \frac{1}{1 - g\delta I_M} g I_M \tilde{p} g \frac{1}{1 + (I_M - \delta I_M)g}$$
$$= W_\delta(\slashed{q}_2, \slashed{q}) - \overline{W}(\slashed{q}_2, \slashed{q}). \quad (F.9)$$

Consider, for example, the term proportional to $\slashed{q}_2$ on the l.h.s. of the last equation. It is equal to

$$\slashed{q}_2 g \frac{1}{1 - \delta I_M g} I_M g \frac{1}{1 + (I_M - \delta I_M)g}$$
$$= \slashed{q}_2 g \frac{1}{1 - \delta I_M g} (1 + (I_M - \delta I_M)g - (1 - \delta I_M g)) \frac{1}{1 + (I_M - \delta I_M)g}$$
$$= \slashed{q}_2 g \frac{1}{1 - \delta I_M g} - \slashed{q}_2 g \frac{1}{1 + (I_M - \delta I_M)g},$$

which equals the $\slashed{q}_2$-dependent term on the r.h.s. of eq. (F.9). In a similar way, the terms proportional to $\slashed{q}$ and the remaining terms can be shown to obey eq. (F.9). Together with eq. (F.8), this confirms eq. (F.6). The proof of eq. (F.5) is thus complete.

Appendix G

Bethe-Salpeter approach with off-shell kernel

In this appendix, we present an alternative solution of the Bethe-Salpeter equation with off-shell pieces in the interaction kernel [140, 141]. In general, the amplitude for a meson-baryon scattering process $B_a(p_a) + \phi_i(q_i) \to B_b(p_b) + \phi_j(q_j)$ can be written as a function $V(t-u, p, \slashed{p}_c - m_c, q_k^2 - M_k^2)$ with masses m_c ($c = a, b$) and M_k ($k = i, j$) for the baryons and mesons, respectively, the total momentum $p = p_a + q_i = p_b + q_j$ and the Mandelstam variables $t = (p_a - p_b)^2$, $u = (p_a - q_j)^2$. It is understood that the off-shell terms $\slashed{p}_a - m_a$ have been moved to the right, whereas the $\slashed{p}_b - m_b$ have been moved to the left. We assume that V is analytic in the variable $t - u$. This is certainly the case for all polynomial interactions which are derived from contact interactions. The Taylor expansion of V in $t - u$ reads

$$V(t-u) = \sum_{i=0}^{\infty} (t-u)^i V_i = V_0 + (t-u)V_1 + (t-u)^2 V_2 + \ldots , \qquad (G.1)$$

where we have suppressed the dependence of V and V_i on $\slashed{p}$ and the off-shell pieces $\slashed{p}_c - m_c$ and $p_k^2 - M_k^2$ for brevity. In the center-of-mass frame the variable $t - u$ is related to the scattering angle θ via

$$t - u = 4|\boldsymbol{p}_a||\boldsymbol{p}_b|\cos\theta + \frac{1}{p^2}(p_a^2 - q_i^2)(p_b^2 - q_j^2), \qquad (G.2)$$

where $|\boldsymbol{p}_{a,b}|$ are the moduli of the c.m. three-momenta. We define a set of orthogonal functions J_l with $l = 0, 1, 2, \ldots$ (or $s, p, d, \ldots$) which are proportional to the Legendre polynomials in the center-of-mass scattering angle with the first few J_l given by

$$\begin{aligned}
J_s &= 1, \\
J_p &= |\boldsymbol{p}_a||\boldsymbol{p}_b|\cos\theta = \frac{1}{4}(t-u) - \frac{1}{4p^2}(p_a^2 - q_i^2)(p_b^2 - q_j^2), \\
J_d &= |\boldsymbol{p}_a|^2|\boldsymbol{p}_b|^2(\cos^2\theta - \frac{1}{3}) = J_p^2 - \frac{1}{3}|\boldsymbol{p}_a|^2|\boldsymbol{p}_b|^2, \\
J_f &= |\boldsymbol{p}_a|^3|\boldsymbol{p}_b|^3(\cos^3\theta - \frac{3}{5}\cos\theta) = J_p^3 - \frac{3}{5}|\boldsymbol{p}_a|^2|\boldsymbol{p}_b|^2 J_p.
\end{aligned} \qquad (G.3)$$

The moduli of the three-momenta in the center-of-mass frame can be expressed in terms of Lorentz scalars as

$$|\boldsymbol{p}_a|^2 = \frac{1}{4p^2}(p^2+q_i^2-p_a^2)^2 - q_i^2,$$
$$|\boldsymbol{p}_b|^2 = \frac{1}{4p^2}(p^2+q_j^2-p_b^2)^2 - q_j^2. \tag{G.4}$$

The expansion of the amplitude V in $t-u$ can now be reformulated as an expansion in J_l, $V = \sum_l V_l J_l$. In order to keep the presentation simple, we restrict ourselves to the Weinberg-Tomozawa interaction from now on. The generalization to more complicated vertex structures is straightforward. For the Weinberg-Tomozawa term V does not depend on $t-u$,

$$V = V_0 = V_s, \tag{G.5}$$

and the division of V into on- and off-shell pieces can be conveniently accomplished by employing the matrix notation (for the reaction $ai \to bj$)

$$V_s^{bj,ai} = \boldsymbol{u}_{bj}^T \check{V}_s^{bj,ai} \boldsymbol{u}_{ai}, \tag{G.6}$$
$$\boldsymbol{u}_{ai}^T(\not{p}_a) = (1, \not{p}_a - m_a),$$
$$\boldsymbol{u}_{bj}^T(\not{p}_b) = (1, \not{p}_b - m_b),$$
$$\check{V}_s^{bj,ai}(\not{p}) = g^{bj,ai} \begin{pmatrix} 2\not{p} - m_a - m_b & -1 \\ -1 & 0 \end{pmatrix}, \tag{G.7}$$

where $g^{bj,ai}$ is the Weinberg-Tomozawa coupling for the channels under consideration as defined in eq. (5.2). The amplitude V is utilized as the interaction kernel in the Bethe-Salpeter equation for the scattering matrix T, cf. eq. (5.5),

$$T(\not{q}_j, \not{q}_i; p) = V(\not{q}_j, \not{q}_i) + \int \frac{d^d l}{(2\pi)^d} V(\not{q}_j, \not{l}) i S(\not{p} - \not{l}) \Delta(l) T(\not{l}, \not{q}_i; p). \tag{G.8}$$

For the decomposition into on- and off-shell pieces we make the same ansatz for the partial wave decomposition of the BSE solution T as for V in eq. (G.6),

$$T_s^{bj,ai} = \boldsymbol{u}_{bj}^T \check{T}_s^{bj,ai} \boldsymbol{u}_{ai}. \tag{G.9}$$

This yields a BSE for $\check{T}_s$ which is a matrix equation in the combined channel and off-shell space

$$\check{T}_s = \check{V}_s + \check{V}_s \check{G}_s \check{T}_s \tag{G.10}$$

with a matrix $\check{G}_s$ which is diagonal in channel space with off-shell submatrices

$$\check{G}_s^{ck,ck} = \int \frac{d^d l}{(2\pi)^d} \boldsymbol{u}_{ck}(\not{l}) i S(\not{l}) \Delta(p-l) \boldsymbol{u}_{ck}^T(\not{l}). \tag{G.11}$$

Hence, the integral expression in the BSE factorizes also without putting the interaction kernel V and the solution T on-shell. In order to solve this equation by matrix inversion, it is most convenient to decompose the matrices $\check{T}_s$, $\check{V}_s$, and $\check{G}_s$ into a Dirac scalar and a piece proportional to $\not{p}$ (see also sect. 5.2).

Appendix H

Loop integrals for the electroproduction amplitude

The tadpole integral is given by

$$I_{\text{M}}^{bj,ai} = \int \frac{d^d l}{(2\pi)^d} \frac{i\delta^{ba}\delta^{ji}}{l^2 - M_j^2 + i0^+} = \left(2M_j^2\bar{\lambda} + \frac{M_j^2}{8\pi^2}\ln\left(\frac{M_j}{\mu}\right)\right)\delta^{ba}\delta^{ji}. \tag{H.1}$$

Here μ is the scale of dimensional regularization and

$$\bar{\lambda} = \frac{\mu^{d-4}}{16\pi^2}\left(\frac{1}{d-4} - \frac{1}{2}[\ln(4\pi) + \Gamma'(1) + 1]\right).$$

In practice, we will use the $\overline{MS}$ renormalization scheme, i.e. terms proportional to $\bar{\lambda}$ will be dropped from all expressions. In the above result, we have neglected terms of $O(4-d)$ since we are interested in the limit $d \to 4$. The diagonal entries in the matrix I_M of eq. (2.2) are given by the above expression for the tadpole integral, where M_j is given by the meson mass of the respective channel. Moreover, the scale μ can vary between the different channels. We shall also define a diagonal matrix I_B with entries given by tadpole integrals where M_j is replaced by the baryon mass m_a of each channel.

Note that we do not make use of the infrared regularization scheme in chapter 5 and in this appendix. The reason is twofold. First, the BSE framework does not allow for a strict renormalization procedure, as we have noted several times, see e.g. sec. 1.5. Here we substitute $\overline{MS}$-renormalized loop integrals for the full loop integrals in all expressions, where the renormalization scales μ for the various channels are treated as adjustable parameters that simulate the contributions from higher-order counterterms. The second reason for suspending the IR scheme here is a more practical one: experience has shown that the use of the IR scheme in the three-flavor case does not necessarily lead to an improvement of the convergence properties of the chiral series (this can of course be traced back to the sizeable kaon masses).

In the following, we will leave out the $+i0^+$ prescription in the loop integrals for brevity. The fundamental scalar loop integral reads

$$I_0(p) = \int \frac{d^d l}{(2\pi)^d} \frac{i}{[(p-l)^2 - m^2][l^2 - M^2]} = \frac{1}{16\pi^2}\Bigg\{-1 + \ln\left(\frac{m^2}{\mu^2}\right) + \frac{M^2 - m^2 + p^2}{2p^2}\ln\left(\frac{M^2}{m^2}\right) - \frac{4|\mathbf{q}|}{\sqrt{p^2}}\operatorname{artanh}\left(\frac{2|\mathbf{q}|\sqrt{p^2}}{(m+M)^2 - p^2}\right)\Bigg\}, \tag{H.2}$$

where
$$|\mathbf{q}| = \frac{\sqrt{(p^2 - (m+M)^2)(p^2 - (m-M)^2)}}{2\sqrt{p^2}} \tag{H.3}$$

is equal to the modulus of the center-of-mass three momentum for a system with particle masses m and M, and p^2 the total invariant energy squared of the system.
It is useful to define a diagonal matrix $I_{\text{MB}}(p)$, with elements given by the above loop integral I_0 of eq. (H.2), where m is the baryon mass m_a and M is the meson mass M_i of the respective channels. Similarly, we will use matrices I_{BB} and I_{MM}, where m and M are both given by either the baryon mass m_a (for I_{BB}) or by M_i (for I_{MM}), respectively. In the same manner we define

$$I^{bj,ai}_{\text{MBB}} = \int \frac{d^d l}{(2\pi)^d} \frac{i\delta^{ba}\delta^{ji}}{[(p-l)^2 - m_a^2][(p_1-l)^2 - m_a^2][l^2 - M_i^2]}. \tag{H.4}$$

An analogous definition applies for I_{MMB}. Only the diagonal elements are needed here, since the coupling of the photon does not alter the meson-baryon channel. The loop integrals occurring in I_{MBB} and I_{MMB} can be expressed in terms of Spence functions [142]. In the course of the calculation, loop integrals with a tensor structure in the integrand are also encountered. For example,

$$\int \frac{d^d l}{(2\pi)^d} \frac{i\delta^{ba}\delta^{ji} l^\mu}{[(p-l)^2 - m_a^2][l^2 - M_i^2]} = p^\mu [I^{(1)}_{\text{MB}}]^{bj,ai} \tag{H.5}$$

with

$$[I^{(1)}_{\text{MB}}]^{bj,ai} = \frac{1}{2p^2}\left[(p^2 + M_i^2 - m_a^2)I^{bj,ai}_{\text{MB}} + I^{bj,ai}_{\text{B}} - I^{bj,ai}_{\text{M}}\right]. \tag{H.6}$$

The last expression is simplified for equal masses in the propagators, so that, e.g., (in matrix notation)

$$\int \frac{d^d l}{(2\pi)^d} i l^\mu \Delta(p-l)\Delta(l) = \frac{p^\mu}{2} I_{\text{MM}}(p). \tag{H.7}$$

We also need

$$\int \frac{d^d l}{(2\pi)^d} i l^\mu l^\nu \Delta(p-l)\Delta(l) = g^{\mu\nu} I^{(a)}_{\text{MM}}(p) + \frac{p^\mu p^\nu}{p^2} I^{(b)}_{\text{MM}}(p), \tag{H.8}$$

where the coefficients of the tensor structures are given by

$$(d-1) I^{(a)}_{\text{MM}}(p) = (M^2 - \frac{1}{4}p^2) I_{\text{MM}}(p) + \frac{1}{2} I_{\text{M}}, \tag{H.9}$$

$$(d-1) I^{(b)}_{\text{MM}}(p) = (\frac{d}{4}p^2 - M^2) I_{\text{MM}}(p) + (\frac{d}{2} - 1) I_{\text{M}}. \tag{H.10}$$

Here we have employed the meson mass matrix

$$M^{bj,ai} = \delta^{ba}\delta^{ji} M_j.$$

defined in analogy to the baryon mass matrix m, and the term p^2 in the last two equations is of course to be understood as being multiplied by the identity matrix in channel space.
The matrix G of eq. (5.10) can be given in terms of scalar loop integrals already defined:

$$G(p) = \frac{\slashed{p}}{2p^2}\left([p^2 - M^2 + m^2] I_{\text{MB}}(p) + I_{\text{M}} - I_{\text{B}}\right) + m I_{\text{MB}}(p). \tag{H.11}$$

Integrals with a vector or tensor structure in the numerator will also be required in the case of three propagators. In particular, two of the three masses in the propagators will be equal, as already mentioned above. In the following, we will consider the case MBB, noting that the other case (MMB) is then given simply by interchanging m and M. First,

$$\int \frac{d^d l}{(2\pi)^d} \frac{i\delta^{ba}\delta^{ji}l^\mu}{[(p-l)^2-m_a^2][(p_1-l)^2-m_a^2][l^2-M_j^2]} = A^{bj,ai}(p_1,p)(p_1+p)^\mu + B^{bj,ai}(p_1,p)(p_1-p)^\mu. \tag{H.12}$$

The channel matrices A and B defined by this equation read

$$A = \frac{1}{2D}\left\{\left(4\bar{M}^2 - \frac{\Delta p^4}{k^2}\right)I_{\text{MBB}} + 2I_{\text{BB}}(k) - \left(1 - \frac{\Delta p^2}{k^2}\right)I_{\text{MB}}(p_1) - \left(1 + \frac{\Delta p^2}{k^2}\right)I_{\text{MB}}(p)\right\}, \tag{H.13}$$

$$B = \frac{\Delta p^2}{2k^2 D}\left\{(4\bar{M}^2 + k^2 - 4\bar{p}^2)I_{\text{MBB}} + 2I_{\text{BB}}(k) - \left(1 - \frac{4\bar{p}^2 - k^2}{\Delta p^2}\right)I_{\text{MB}}(p_1)\right.$$
$$\left.- \left(1 + \frac{4\bar{p}^2 - k^2}{\Delta p^2}\right)I_{\text{MB}}(p)\right\}. \tag{H.14}$$

Here we have used the following notation:

$$\bar{p}^2 = \frac{p_1^2 + p^2}{2}, \tag{H.15}$$

$$\bar{M}^2 = \frac{1}{2}(\bar{p}^2 + M^2 - m^2), \tag{H.16}$$

$$\Delta p^2 = p^2 - p_1^2, \tag{H.17}$$

$$k = p - p_1, \tag{H.18}$$

$$D = 4\bar{p}^2 - k^2 - \frac{\Delta p^4}{k^2}, \tag{H.19}$$

and the abbreviation $\Delta p^4 \equiv (\Delta p^2)^2$. The loop integral with three propagators and tensor structure is given by

$$I_\Delta^{\mu\nu} = \int \frac{d^d l}{(2\pi)^d} \frac{i\delta^{ba}\delta^{ji}l^\mu l^\nu}{[(p-l)^2-m_a^2][(p_1-l)^2-m_a^2][l^2-M_j^2]} = C_1^{bj,ai}g^{\mu\nu} + C_2^{bj,ai}(p_1+p)^\mu(p_1+p)^\nu$$
$$+ C_3^{bj,ai}(p_1-p)^\mu(p_1-p)^\nu + C_4^{bj,ai}((p_1+p)^\mu(p_1-p)^\nu + (p_1+p)^\nu(p_1-p)^\mu).$$

The coefficient matrices C_i read as follows:

$$C_1 = \frac{1}{d-2}\left\{M^2 I_{\text{MBB}} + \frac{1}{2}I_{\text{BB}}(k) - 2\bar{M}^2 A + \frac{\Delta p^2}{2}B\right\}, \tag{H.20}$$

$$C_2 = \frac{1}{k^2 D}\left\{k^2(M^2 I_{\text{MBB}} + I_{\text{BB}}(k)) + \frac{\Delta p^2}{2}(k^2 B - \Delta p^2 A)\right.$$
$$\left.- \frac{k^2}{4}(I_{\text{MB}}^{(1)}(p_1) + I_{\text{MB}}^{(1)}(p)) + \frac{\Delta p^2}{4}(I_{\text{MB}}^{(1)}(p_1) - I_{\text{MB}}^{(1)}(p)) - (d-1)k^2 C_1\right\}, \tag{H.21}$$

$$C_3 = \frac{1}{k^2 D}\left\{(4\bar{p}^2 - k^2)(M^2 I_{\text{MBB}} + \frac{1}{2}I_{\text{BB}}(k)) - 2\bar{M}^2((4\bar{p}^2 - k^2)A - \Delta p^2 B)\right.$$
$$\left. + \frac{1}{4}(4\bar{p}^2 - \Delta p^2 - k^2)I_{\text{MB}}^{(1)}(p_1) + \frac{1}{4}(4\bar{p}^2 + \Delta p^2 - k^2)I_{\text{MB}}^{(1)}(p) - (d-1)(4\bar{p}^2 - k^2)C_1\right\}, \quad \text{(H.22)}$$

$$C_4 = \frac{1}{k^2 D}\left\{\Delta p^2(M^2 I_{\text{MBB}} + I_{\text{BB}}(k)) - \frac{\Delta p^2}{2}((4\bar{p}^2 - k^2)A - \Delta p^2 B)\right.$$
$$\left. + \frac{1}{4}(4\bar{p}^2 - \Delta p^2 - k^2)I_{\text{MB}}^{(1)}(p_1) - \frac{1}{4}(4\bar{p}^2 + \Delta p^2 - k^2)I_{\text{MB}}^{(1)}(p) - (d-1)\Delta p^2 C_1\right\}. \quad \text{(H.23)}$$

A and B are given in eqs. (H.13) and (H.14), while $I_{\text{MB}}^{(1)}$ is defined in eq. (H.6). Note that in the limit $d \to 4$, C_1 will aquire an additional finite contribution due to the divergent terms in the loop integrals. In particular,

$$C_1 \to C_1(d=4) - \frac{1}{64\pi^2}, \quad \text{(H.24)}$$

$$(d-1)C_1 \to 3C_1(d=4) + \frac{1}{32\pi^2}. \quad \text{(H.25)}$$

As already mentioned, the results for the MMB case can be obtained from the above formulae by interchanging $m \leftrightarrow M$. The corresponding coefficients will be denoted by $\tilde{A}, \tilde{B}$ and $\tilde{C}_i$. Finally, the preceding results can be used to derive the decompositions

$$I_F = \int \frac{d^d l}{(2\pi)^d} iS(\slashed{p} - \slashed{l})eQ_B\gamma^\mu S(\slashed{p}_1 - \slashed{l})\Delta(l)$$
$$= \gamma^\mu F_1 + \slashed{p}\gamma^\mu F_2 + \gamma^\mu \slashed{p}_1 F_3 + \slashed{p}\gamma^\mu \slashed{p}_1 F_4 + p^\mu F_5 + p_1^\mu F_6 + p^\mu \slashed{p} F_7 + p^\mu \slashed{p}_1 F_8 + p_1^\mu \slashed{p} F_9 + p_1^\mu \slashed{p}_1 F_{10} \quad \text{(H.26)}$$

and

$$I_{\tilde{F}} = \int \frac{d^d l}{(2\pi)^d} i\Delta(p-l)eQ_M(p+p_1-2l)^\mu \Delta(p_1-l)S(\slashed{l})$$
$$= \gamma^\mu \tilde{F}_1 + \slashed{p}\gamma^\mu \tilde{F}_2 + \gamma^\mu \slashed{p}_1 \tilde{F}_3 + \slashed{p}\gamma^\mu \slashed{p}_1 \tilde{F}_4 + p^\mu \tilde{F}_5 + p_1^\mu \tilde{F}_6 + p^\mu \slashed{p} \tilde{F}_7 + p^\mu \slashed{p}_1 \tilde{F}_8 + p_1^\mu \slashed{p} \tilde{F}_9 + p_1^\mu \slashed{p}_1 \tilde{F}_{10}, \quad \text{(H.27)}$$

where the coefficients of the Lorentz structures are given by

$$F_1 = eQ_B(2C_1 + (m^2 - M^2)I_{\text{MBB}} - I_{\text{BB}}(k) + p_1^2(A+B) + p^2(A-B)),$$
$$F_2 = eQ_B m I_{\text{MBB}},$$
$$F_3 = F_2,$$
$$F_4 = eQ_B(I_{\text{MBB}} - 2A),$$
$$F_5 = 2eQ_B m(B-A),$$
$$F_6 = -2eQ_B m(B+A),$$
$$F_7 = 2eQ_B(B - A + C_2 + C_3 - 2C_4),$$
$$F_8 = 2eQ_B(C_2 - C_3),$$
$$F_9 = F_8,$$

$$F_{10} = 2eQ_{\text{B}}(C_2 + C_3 + 2C_4 - A - B),$$
$$\tilde{F}_1 = -2eQ_{\text{M}}\tilde{C}_1,$$
$$\tilde{F}_2 = \tilde{F}_3 = \tilde{F}_4 = 0,$$
$$\tilde{F}_5 = eQ_{\text{M}}(mI_{\text{MMB}} + 2m(\tilde{B} - \tilde{A})),$$
$$\tilde{F}_6 = eQ_{\text{M}}(mI_{\text{MMB}} - 2m(\tilde{B} + \tilde{A})),$$
$$\tilde{F}_7 = eQ_{\text{M}}(\tilde{A} - \tilde{B} - 2(\tilde{C}_2 + \tilde{C}_3 - 2\tilde{C}_4)),$$
$$\tilde{F}_8 = eQ_{\text{M}}(\tilde{A} + \tilde{B} - 2(\tilde{C}_2 - \tilde{C}_3)),$$
$$\tilde{F}_9 = eQ_{\text{M}}(\tilde{A} - \tilde{B} - 2(\tilde{C}_2 - \tilde{C}_3)),$$
$$\tilde{F}_{10} = eQ_{\text{M}}(\tilde{A} + \tilde{B} - 2(\tilde{C}_2 + \tilde{C}_3 + 2\tilde{C}_4)).$$

The following relations are helpful, e.g. when checking gauge invariance:

$$\begin{aligned}G_1(p) &= \bar{p}^2 F_4 - \frac{\Delta p^2}{2}(F_4 + F_7 + F_9) + \frac{k^2}{2}(F_9 - F_7) - F_1 \\ &= \frac{k^2}{2}(\tilde{F}_9 - \tilde{F}_7) - \frac{\Delta p^2}{2}(\tilde{F}_9 + \tilde{F}_7) - \tilde{F}_1,\\ G_1(p_1) &= \bar{p}^2 F_4 + \frac{\Delta p^2}{2}(F_4 + F_8 + F_{10}) + \frac{k^2}{2}(F_8 - F_{10}) - F_1 \\ &= \frac{k^2}{2}(\tilde{F}_8 - \tilde{F}_{10}) + \frac{\Delta p^2}{2}(\tilde{F}_8 + \tilde{F}_{10}) - \tilde{F}_1,\end{aligned}$$

and also

$$\begin{aligned}G_0(p) - G_0(p_1) &= \frac{k^2}{2}(F_6 - F_5) - \frac{\Delta p^2}{2}(2F_2 + F_5 + F_6) \\ &= \frac{k^2}{2}(\tilde{F}_6 - \tilde{F}_5) - \frac{\Delta p^2}{2}(\tilde{F}_5 + \tilde{F}_6).\end{aligned}$$

As a further check of the calculation we have utilized the routines provided by the FeynCalc package [143].

Appendix I

Decomposition of the electroproduction amplitude

We start with the amplitudes of eqs. (5.42)-(5.44).

$$S_s^\mu = (\slashed{q}\slashed{p}S_s^{\slashed{q}\slashed{p}\gamma} + \slashed{q}S_s^{\slashed{q}\gamma} + \slashed{p}S_s^{\slashed{p}\gamma} + S_s^\gamma)\gamma^\mu\gamma_5 \qquad (\text{I.1})$$

with

$$S_s^{\slashed{q}\slashed{p}\gamma} = \frac{e}{m_p^2 - s}(\Gamma_2(p) - m_p\Gamma_1(p)),$$

$$S_s^{\slashed{q}\gamma} = \frac{e}{m_p^2 - s}(s\Gamma_1(p) - m_p\Gamma_2(p)),$$

$$S_s^{\slashed{p}\gamma} = \frac{e}{m_p^2 - s}(\Gamma_4(p) - m_p\Gamma_3(p)),$$

$$S_s^\gamma = \frac{e}{m_p^2 - s}(s\Gamma_3(p) - m_p\Gamma_4(p)).$$

Here, m_p is the proton mass and $s \equiv p^2 = (p_1 + k)^2$. Moreover,

$$S_u^\mu = q^\mu\gamma_5 S_u^q + \slashed{q}\gamma^\mu\gamma_5 S_u^{\slashed{q}\gamma} + \gamma^\mu\gamma_5 S_u^\gamma \qquad (\text{I.2})$$

with

$$S_u^q = -2S_u^{\slashed{q}\gamma} = \frac{2eQ_B}{u - m^2}Y_3,$$

$$S_u^\gamma = \frac{eQ_B}{m^2 - u}\left(\Gamma_1(p_1)m_p(u - m_p^2) - \Gamma_2(p_1)(u - m_p^2) + (m_p^2 - m_p m)\Gamma_3(p_1) + (m - m_p)\Gamma_4(p_1)\right),$$

and

$$S_t^\mu = \left(p^\mu S_t^p + p_1^\mu S_t^{p_1} + q^\mu S_t^q + \slashed{q}q^\mu S_t^{\slashed{q}q} + \slashed{p}q^\mu S_t^{\slashed{p}q} + \slashed{q}p^\mu S_t^{\slashed{q}p} + \slashed{p}p^\mu S_t^{\slashed{p}p} + \slashed{q}p_1^\mu S_t^{\slashed{q}p_1} + \slashed{p}p_1^\mu S_t^{\slashed{p}p_1}\right)\gamma_5, \qquad (\text{I.3})$$

with

$$S_t^p = \frac{eQ_M}{t - M^2}Y_1, \qquad S_t^{p_1} = -S_t^p, \qquad S_t^q = -2S_t^p,$$

$$S_t^{\slashed{q}q} = -\frac{2eQ_M}{t - M^2}Y_2, \qquad S_t^{\slashed{p}q} = -S_t^{\slashed{q}q}, \qquad S_t^{\slashed{q}p} = \frac{eQ_M}{t - M^2}Y_2,$$

$$S_t^{\slashed{p}p} = -S_t^{\slashed{q}p}, \qquad S_t^{\slashed{q}p_1} = -S_t^{\slashed{q}p}, \qquad S_t^{\slashed{p}p_1} = +S_t^{\slashed{q}p}.$$

The foregoing results deserve some comments. We use the abbreviations

$$Y_1 = m_p^2 \Gamma_1(p_1) - m_p(\Gamma_2(p_1) + \Gamma_3(p_1)) + \Gamma_4(p_1), \tag{I.4}$$

$$Y_2 = \Gamma_2(p_1) - m_p \Gamma_1(p_1), \tag{I.5}$$

$$Y_3 = (m_p^2 + m_p m)\Gamma_1(p_1) - (m + m_p)\Gamma_2(p_1) - m_p \Gamma_3(p_1) + \Gamma_4(p_1). \tag{I.6}$$

Furthermore, we have anticipated that the full amplitude will be multiplied by spinors, so we set $\not{p}_1 \gamma_5 \to -m_p \gamma_5$. The Mandelstam variables u and t are understood to be diagonal matrices in channel space and are defined via the center-of-mass (c.m.) relations

$$t = M^2 + k^2 - 2E_k E_q + 2|\mathbf{k}||\mathbf{q}|\cos\theta, \tag{I.7}$$

$$u = k^2 + m_p^2 + m^2 + M^2 - s - t, \tag{I.8}$$

where $|\mathbf{q}|$ and $|\mathbf{k}|$ are the moduli of the c.m. three-momenta of the outgoing meson and the incoming photon, respectively (see eq. (H.3)), θ is the scattering angle in the c.m. frame, and the c.m. energies are given by

$$E_k = \sqrt{|\mathbf{k}|^2 + k^2},$$
$$E_q = \sqrt{|\mathbf{q}|^2 + M^2}.$$

Of course, multiplication of scalar quantities (such as k^2 or m_p) by the unit matrix in channel space is implied where necessary and fractions involving matrix-valued denominators denote matrix inversions.

Now we come to the graphs where the photon couples to internal meson or baryon lines. The decomposition of S_B^μ reads

$$S_B^\mu = (\gamma^\mu S_B^\gamma + p^\mu S_B^p + p_1^\mu S_B^{p_1} + \not{q}\gamma^\mu S_B^{\not{q}\gamma} + \not{p}\gamma^\mu S_B^{\not{p}\gamma} + \not{q}\not{p}\gamma^\mu S_B^{\not{q}\not{p}\gamma} + \not{q}p^\mu S_B^{\not{q}p} + \not{p}p^\mu S_B^{\not{p}p}$$
$$+ \not{q}\not{p}p^\mu S_B^{\not{q}\not{p}p} + \not{q}p_1^\mu S_B^{\not{q}p_1} + \not{p}p_1^\mu S_B^{\not{p}p_1} + \not{q}\not{p}p_1^\mu S_B^{\not{q}\not{p}p_1})\gamma_5 \tag{I.9}$$

with

$$S_B^\gamma = m_p[T_8 F_3 - T_5 m F_3 - T_5 eQ_B G_1(p_1) + s(T_3 F_3 + T_5 F_4 + T_7 F_4 - T_3 m F_4)]Y_3$$
$$- T_8 F_1 Y_3 + T_5 m F_1 Y_3 + T_5 eQ_B G_0(p_1) Y_3$$
$$- T_5 eQ_B I_M Y_2 + T_8 G_0(p) eQ_B Y_2 - T_5 m G_0(p) eQ_B Y_2$$
$$- s(T_3 F_1 Y_3 + T_5 F_2 Y_3 + T_7 F_2 Y_3 - T_3 G_0(p) eQ_B Y_2 - T_3 m F_2 Y_3 - T_5 G_1(p) eQ_B Y_2$$
$$- T_7 G_1(p) eQ_B Y_2 + T_3 m G_1(p) eQ_B Y_2),$$

$$S_B^p = m_p[T_8 F_8 Y_3 - T_5 m F_8 Y_3 + sT_3 F_8 Y_3] - T_8 F_5 Y_3 + T_5 m F_5 Y_3$$
$$- s(T_3 F_5 Y_3 + T_5 F_7 Y_3 + T_7 F_7 Y_3 - T_3 m F_7 Y_3),$$

$$S_B^{p_1} = m_p[T_8 F_{10} Y_3 - T_5 m F_{10} Y_3 + sT_3 F_{10} Y_3] - T_8 F_6 Y_3 + T_5 m F_6 Y_3$$
$$- s(T_3 F_6 + T_5 F_9 + T_7 F_9 - T_3 m F_9)Y_3,$$

$$S_B^{\not{q}\gamma} = m_p[T_6 F_3 - T_2 m F_3 - T_2 eQ_B G_1(p_1) + s(T_1 F_3 + T_2 F_4 + T_4 F_4 - T_1 m F_4)]Y_3$$
$$- T_6 F_1 Y_3 + T_2 m F_1 Y_3 + T_2 eQ_B G_0(p_1) Y_3$$
$$- T_2 eQ_B I_M Y_2 + T_6 G_0(p) eQ_B Y_2 - T_2 m G_0(p) eQ_B Y_2$$
$$- s(T_1 F_1 Y_3 + T_2 F_2 Y_3 + T_4 F_2 Y_3 - T_1 G_0(p) eQ_B Y_2 - T_1 m F_2 Y_3 - T_2 G_1(p) eQ_B Y_2$$
$$- T_4 G_1(p) eQ_B Y_2 + T_1 m G_1(p) eQ_B Y_2),$$

$$\begin{aligned}
S_{\mathrm{B}}^{\slashed{p}\gamma} &= m_p[T_5F_3 + T_7F_3 + T_8F_4 - T_3mF_3 - T_3eQ_{\mathrm{B}}G_1(p_1) - T_5mF_4 + sT_3F_4]Y_3 \\
&\quad - T_5F_1Y_3 - T_7F_1Y_3 - T_8F_2Y_3 + T_3mF_1Y_3 + T_3eQ_{\mathrm{B}}G_0(p_1)Y_3 - T_3eQ_{\mathrm{B}}I_{\mathrm{M}}Y_2 \\
&\quad + T_5G_0(p)eQ_{\mathrm{B}}Y_2 + T_5mF_2Y_3 + T_7G_0(p)eQ_{\mathrm{B}}Y_2 + T_8G_1(p)eQ_{\mathrm{B}}Y_2 - T_3mG_0(p)eQ_{\mathrm{B}}Y_2 \\
&\quad - T_5mG_1(p)eQ_{\mathrm{B}}Y_2 + s(T_3G_1(p)eQ_{\mathrm{B}}Y_2 - T_3F_2Y_3)\,, \\
S_{\mathrm{B}}^{\slashed{q}\slashed{p}\gamma} &= m_p[T_2F_3 + T_4F_3 + T_6F_4 - T_1mF_3 - T_1eQ_{\mathrm{B}}G_1(p_1) - T_2mF_4 + sT_1F_4]Y_3 \\
&\quad - T_2F_1Y_3 - T_4F_1Y_3 - T_6F_2Y_3 + T_1mF_1Y_3 + T_1eQ_{\mathrm{B}}G_0(p_1)Y_3 - T_1eQ_{\mathrm{B}}I_{\mathrm{M}}Y_2 \\
&\quad + T_2G_0(p)eQ_{\mathrm{B}}Y_2 + T_2mF_2Y_3 + T_4G_0(p)eQ_{\mathrm{B}}Y_2 + T_6G_1(p)eQ_{\mathrm{B}}Y_2 - T_1mG_0(p)eQ_{\mathrm{B}}Y_2 \\
&\quad - T_2mG_1(p)eQ_{\mathrm{B}}Y_2 - s(T_1F_2Y_3 - T_1G_1(p)eQ_{\mathrm{B}}Y_2)\,, \\
S_{\mathrm{B}}^{\slashed{q}p} &= m_p[T_6F_8 - T_2mF_8 + sT_1F_8]Y_3 - T_6F_5Y_3 + T_2mF_5Y_3 \\
&\quad - s(T_1F_5 + T_2F_7 + T_4F_7 - T_1mF_7)Y_3\,, \\
S_{\mathrm{B}}^{\slashed{p}p} &= m_p[T_5F_8 + T_7F_8 - T_3mF_8]Y_3 - (T_5F_5 + T_7F_5 + T_8F_7 - T_3mF_5 \\
&\quad - T_5mF_7 + sT_3F_7)Y_3\,, \\
S_{\mathrm{B}}^{\slashed{q}\slashed{p}p} &= m_p[T_2F_8 + T_4F_8 - T_1mF_8]Y_3 \\
&\quad - (T_2F_5 + T_4F_5 + T_6F_7 - T_1mF_5 \\
&\quad - T_2mF_7 + sT_1F_7)Y_3\,, \\
S_{\mathrm{B}}^{\slashed{q}p_1} &= m_p[T_6F_{10} - T_2mF_{10} + sT_1F_{10}]Y_3 \\
&\quad - T_6F_6Y_3 + T_2mF_6Y_3 - s(T_1F_6 + T_2F_9 + T_4F_9 - T_1mF_9)Y_3\,, \\
S_{\mathrm{B}}^{\slashed{p}p_1} &= m_p[T_5F_{10} + T_7F_{10} - T_3mF_{10}]Y_3 - (T_5F_6 + T_7F_6 + T_8F_9 - T_3mF_6 \\
&\quad - T_5mF_9 + sT_3F_9)Y_3\,, \\
S_{\mathrm{B}}^{\slashed{q}\slashed{p}p_1} &= m_p[T_2F_{10} + T_4F_{10} - T_1mF_{10}]Y_3 - (T_2F_6 + T_4F_6 + T_6F_9 - T_1mF_6 \\
&\quad - T_2mF_9 + sT_1F_9)Y_3\,,
\end{aligned}$$

while the decomposition of S_{M}^μ reads

$$S_{\mathrm{M}}^\mu = \big(\gamma^\mu S_{\mathrm{M}}^\gamma + p^\mu S_{\mathrm{M}}^p + p_1^\mu S_{\mathrm{M}}^{p_1} + \slashed{q}\gamma^\mu S_{\mathrm{M}}^{\slashed{q}\gamma} + \slashed{p}\gamma^\mu S_{\mathrm{M}}^{\slashed{p}\gamma} + \slashed{q}\slashed{p}\gamma^\mu S_{\mathrm{M}}^{\slashed{q}\slashed{p}\gamma} + \slashed{q}p^\mu S_{\mathrm{M}}^{\slashed{q}p} + \slashed{p}p^\mu S_{\mathrm{M}}^{\slashed{p}p} + \slashed{q}\slashed{p}p^\mu S_{\mathrm{M}}^{\slashed{q}\slashed{p}p} \\ + \slashed{q}p_1^\mu S_{\mathrm{M}}^{\slashed{q}p_1} + \slashed{p}p_1^\mu S_{\mathrm{M}}^{\slashed{p}p_1} + \slashed{q}\slashed{p}p_1^\mu S_{\mathrm{M}}^{\slashed{q}\slashed{p}p_1}\big)\gamma_5\,, \tag{I.10}$$

with

$$\begin{aligned}
S_{\mathrm{M}}^\gamma &= T_5m\tilde{F}_1Y_3 - T_8\tilde{F}_1Y_3 - T_5eQ_{\mathrm{M}}d_1Y_2 - sT_3\tilde{F}_1Y_3\,, \\
S_{\mathrm{M}}^p &= m_p[T_8\tilde{F}_8Y_3 - T_5m\tilde{F}_8Y_3 - T_5eQ_{\mathrm{M}}d_2Y_2 + sT_3\tilde{F}_8Y_3] - T_8\tilde{F}_5Y_3 + T_5m\tilde{F}_5Y_3 \\
&\quad - s(T_3\tilde{F}_5Y_3 + T_5\tilde{F}_7Y_3 + T_7\tilde{F}_7Y_3 - T_3m\tilde{F}_7Y_3 + T_3eQ_{\mathrm{M}}d_2Y_2)\,, \\
S_{\mathrm{M}}^{p_1} &= m_p[T_8\tilde{F}_{10}Y_3 - T_5m\tilde{F}_{10}Y_3 + T_5eQ_{\mathrm{M}}d_2Y_2 + sT_3\tilde{F}_{10}Y_3] - T_8\tilde{F}_6Y_3 + T_5m\tilde{F}_6Y_3 \\
&\quad - s(T_3\tilde{F}_6 + T_5\tilde{F}_9 + T_7\tilde{F}_9 - T_3m\tilde{F}_9)Y_3 + sT_3eQ_{\mathrm{M}}d_2Y_2\,, \\
S_{\mathrm{M}}^{\slashed{q}\gamma} &= T_2m\tilde{F}_1Y_3 - T_6\tilde{F}_1Y_3 - sT_1\tilde{F}_1Y_3 - T_2eQ_{\mathrm{M}}d_1Y_2\,, \\
S_{\mathrm{M}}^{\slashed{p}\gamma} &= T_3m\tilde{F}_1Y_3 - T_5\tilde{F}_1Y_3 - T_7\tilde{F}_1Y_3 - T_3eQ_{\mathrm{M}}d_1Y_2\,, \\
S_{\mathrm{M}}^{\slashed{q}\slashed{p}\gamma} &= T_1m\tilde{F}_1Y_3 - T_2\tilde{F}_1Y_3 - T_4\tilde{F}_1Y_3 - T_1eQ_{\mathrm{M}}d_1Y_2\,, \\
S_{\mathrm{M}}^{\slashed{q}p} &= m_p[T_6\tilde{F}_8Y_3 - T_2m\tilde{F}_8Y_3 + sT_1\tilde{F}_8Y_3 - T_2eQ_{\mathrm{M}}d_2Y_2] - T_6\tilde{F}_5Y_3 + T_2m\tilde{F}_5Y_3 \\
&\quad - s(T_1\tilde{F}_5 + T_2\tilde{F}_7 + T_4\tilde{F}_7 - T_1m\tilde{F}_7)Y_3 - sT_1eQ_{\mathrm{M}}d_2Y_2\,,
\end{aligned}$$

$$\begin{aligned}
S_M^{\not p p} &= m_p[T_5\tilde{F}_8 Y_3 + T_7\tilde{F}_8 Y_3 - T_3 m\tilde{F}_8 Y_3 - T_3 eQ_M d_2 Y_2]\\
&\quad - (T_5\tilde{F}_5 + T_7\tilde{F}_5 + T_8\tilde{F}_7 - T_3 m\tilde{F}_5 - T_5 m\tilde{F}_7)Y_3 - sT_3\tilde{F}_7 Y_3 - T_5 eQ_M d_2 Y_2\,,\\
S_M^{\not q \not p p} &= m_p[T_2\tilde{F}_8 Y_3 + T_4\tilde{F}_8 Y_3 - T_1 m\tilde{F}_8 Y_3 - T_1 eQ_M d_2 Y_2]\\
&\quad - (T_2\tilde{F}_5 + T_4\tilde{F}_5 + T_6\tilde{F}_7 - T_1 m\tilde{F}_5 - T_2 m\tilde{F}_7) - sT_1\tilde{F}_7 Y_3 - T_2 eQ_M d_2 Y_2,\\
S_M^{\not q p 1} &= m_p[T_6\tilde{F}_{10}Y_3 - T_2 m\tilde{F}_{10}Y_3 + sT_1\tilde{F}_{10}Y_3 + T_2 eQ_M d_2 Y_2] - T_6\tilde{F}_6 Y_3 + T_2 m\tilde{F}_6 Y_3\\
&\quad + sT_1 eQ_M d_2 Y_2 - s(T_1\tilde{F}_6 + T_2\tilde{F}_9 + T_4\tilde{F}_9 - T_1 m\tilde{F}_9)Y_3\,,\\
S_M^{\not p p 1} &= m_p[T_5\tilde{F}_{10}Y_3 + T_7\tilde{F}_{10}Y_3 - T_3 m\tilde{F}_{10}Y_3 + T_3 eQ_M d_2 Y_2]\\
&\quad - (T_5\tilde{F}_6 + T_7\tilde{F}_6 + T_8\tilde{F}_9 - T_3 m\tilde{F}_6 - T_5 m\tilde{F}_9)Y_3 - sT_3\tilde{F}_9 Y_3 + T_5 eQ_M d_2 Y_2\,,\\
S_M^{\not q \not p p 1} &= m_p[T_2\tilde{F}_{10}Y_3 + T_4\tilde{F}_{10}Y_3 - T_1 m\tilde{F}_{10}Y_3 + T_1 eQ_M d_2 Y_2]\\
&\quad - (T_2\tilde{F}_6 + T_4\tilde{F}_6 + T_6\tilde{F}_9 - T_1 m\tilde{F}_6 - T_2 m\tilde{F}_9)Y_3\\
&\quad - sT_1\tilde{F}_9 Y_3 + T_2 eQ_M d_2 Y_2\,.
\end{aligned}$$

Here we have used the abbreviations $T_i \equiv T_i(p)$ and

$$d_1 = -2\left[\frac{1}{3}\left((M^2 - \frac{1}{4}k^2)I_{MM}(k) + \frac{1}{2}I_M\right) + \delta\right],$$

$$d_2 = \frac{1}{2}I_{MM}(k) - \frac{2}{k^2}\left[\frac{1}{3}\left((k^2 - M^2)I_{MM}(k) + I_M\right) - \delta\right],$$

$$\delta = \frac{1}{48\pi^2}\left[\frac{1}{6}k^2 - M^2\right].$$

The $T_i(p)$ are defined in eq. (5.28), while the F_i and $\tilde{F}_i$ are defined in eq. (H.26) and (H.27), respectively. The term δ stems from the limit $d \to 4$ in eqs. (H.9) and (H.10).

Next we turn to the class of diagrams derived from the 'Kroll-Ruderman'-term (called 'Class 4' in sect. 5.3):

$$S_{KR}^\mu = (S_{KR}^\gamma + \not q S_{KR}^{\not q \gamma} + \not p S_{KR}^{\not p \gamma} + \not q \not p S_{KR}^{\not q \not p \gamma})\gamma^\mu \gamma_5 \tag{I.11}$$

with

$$\begin{aligned}
S_{KR}^\gamma &= eQ_M\hat{g} + [(T_8 - T_5 m + sT_3)G_0(p) + s(T_5 + T_7 - T_3 m)G_1(p) - T_5 I_M]eQ_M\hat{g}\,,\\
S_{KR}^{\not q \gamma} &= [(T_6 - T_2 m + sT_1)G_0(p) + s(T_2 + T_4 - T_1 m)G_1(p) - T_2 I_M]eQ_M\hat{g}\,,\\
S_{KR}^{\not p \gamma} &= [(T_5 + T_7 - T_3 m)G_0(p) + (sT_3 + T_8 - T_5 m)G_1(p) - T_3 I_M]eQ_M\hat{g}\,,\\
S_{KR}^{\not q \not p \gamma} &= [(T_2 + T_4 - T_1 m)G_0(p) + (sT_1 + T_6 - T_2 m)G_1(p) - T_1 I_M]eQ_M\hat{g}\,.
\end{aligned}$$

Here, the first term in the first line is the contribution from the tree graph, and $T_i \equiv T_i(p)$. The last class of graphs is $S_{WT1}^\mu + S_{WT2}^\mu$, where

$$S_{WT1}^\mu = \gamma^\mu \gamma_5 S_{WT1}^\gamma \tag{I.12}$$

with

$$S_{WT1}^\gamma = (Q_M g + gQ_M)((G_0(p_1) - m_p G_1(p_1))Y_3 - I_M Y_2)\,,$$

and

$$S_{WT2}^\mu = (S_{WT2}^\gamma + \not q S_{WT2}^{\not q \gamma} + \not p S_{WT2}^{\not p \gamma} + \not q \not p S_{WT2}^{\not q \not p \gamma})\gamma^\mu \gamma_5 \tag{I.13}$$

with

$$S^\gamma_{WT2} = [s(T_3 G_0(p) + (T_5 + T_7 - T_3 m)G_1(p)) + (T_8 - T_5 m)G_0(p) - T_5 I_M]S^\gamma_{WT1},$$
$$S^{q\gamma}_{WT2} = [s(T_1 G_0(p) + (T_2 + T_4 - T_1 m)G_1(p)) + (T_6 - T_2 m)G_0(p) - T_2 I_M]S^\gamma_{WT1},$$
$$S^{\not p\gamma}_{WT2} = [(sT_3 - T_5 m + T_8)G_1(p) + (T_5 + T_7 - T_3 m)G_0(p) - T_3 I_M]S^\gamma_{WT1},$$
$$S^{\not q \not p \gamma}_{WT2} = [(sT_1 - T_2 m + T_6)G_1(p) + (T_2 + T_4 - T_1 m)G_0(p) - T_1 I_M]S^\gamma_{WT1}.$$

Adding the contributions of eqs. (I.1)-(I.3) and eqs. (I.9)-(I.13), we can decompose the full photoproduction amplitude $\mathcal{M}^\mu$ (see eq. (5.53)) as follows:

$$\begin{aligned}\mathcal{M}^\mu &= \gamma^\mu \gamma_5 \mathcal{M}_1 + q^\mu \gamma_5 \mathcal{M}_2 + p^\mu \gamma_5 \mathcal{M}_3 + p_1^\mu \gamma_5 \mathcal{M}_4 + \not q \gamma^\mu \gamma_5 \mathcal{M}_5 + \not p \gamma^\mu \gamma_5 \mathcal{M}_6 + \not q \not p \gamma^\mu \gamma_5 \mathcal{M}_7 \\ &+ \not q q^\mu \gamma_5 \mathcal{M}_8 + \not p q^\mu \gamma_5 \mathcal{M}_9 + \not q p^\mu \gamma_5 \mathcal{M}_{10} + \not p p^\mu \gamma_5 \mathcal{M}_{11} + \not q \not p p^\mu \gamma_5 \mathcal{M}_{12} + \not q p_1^\mu \gamma_5 \mathcal{M}_{13} \\ &+ \not p p_1^\mu \gamma_5 \mathcal{M}_{14} + \not q \not p p_1^\mu \gamma_5 \mathcal{M}_{15}\,.\end{aligned} \quad (I.14)$$

This can be simplified, using the Dirac equation and momentum conservation, to arrive at the operator basis given by the $\mathcal{N}^\mu_k$ commonly used for photoproduction processes, see eq. (J.1). The relation between the corresponding coefficients B_k and the functions $\mathcal{M}_j$ used in eq. (I.14) is given in eq. (J.2). The decomposition of the amplitudes into the various Dirac structures is obtained by employing the FeynCalc package [143].

Appendix J

Invariant amplitudes for the electroproduction process

We consider the reaction

$$\gamma(k) + p(p_1, m_1) \to B(p_2, m_2) + M(q, M_\phi)$$

and define the Mandelstam variables as usual,

$$s = (p_1 + k)^2, \quad u = (p_1 - q)^2, \quad t = (p_2 - p_1)^2.$$

They obey the constraint

$$s + t + u = m_1^2 + m_2^2 + M_\phi^2 + k^2.$$

The amplitude can be decomposed as

$$T_{fi} = i\epsilon_\mu \bar{u}_2 \sum_{k=1}^{8} B_k \mathcal{N}_k^\mu u_1, \qquad (J.1)$$

where the operator basis is given by

$$\mathcal{N}_1^\mu = \gamma_5 \gamma^\mu \slashed{k}, \qquad \mathcal{N}_2^\mu = 2\gamma_5 P^\mu, \qquad \mathcal{N}_3^\mu = 2\gamma_5 q^\mu,$$
$$\mathcal{N}_4^\mu = 2\gamma_5 k^\mu, \qquad \mathcal{N}_5^\mu = \gamma_5 \gamma^\mu, \qquad \mathcal{N}_6^\mu = \gamma_5 \slashed{k} P^\mu,$$
$$\mathcal{N}_7^\mu = \gamma_5 \slashed{k} k^\mu, \qquad \mathcal{N}_8^\mu = \gamma_5 \slashed{k} q^\mu.$$

Here, $P = \frac{1}{2}(p_1 + p_2)$. The relation to the coefficient functions $\mathcal{M}_j$ used in eq. (I.14) is

$$B_1 = -\mathcal{M}_5 - \mathcal{M}_6 + m_2 \mathcal{M}_7,$$
$$B_2 = \frac{1}{2}\mathcal{M}_3 + \frac{1}{2}\mathcal{M}_4 + \mathcal{M}_5 + \mathcal{M}_6 - m_2 \mathcal{M}_7$$
$$- \frac{1}{2}(m_1 + m_2)\mathcal{M}_{10} - \frac{1}{2}m_1 \mathcal{M}_{11} + \frac{1}{2}(s + m_1 m_2)\mathcal{M}_{12}$$
$$- \frac{1}{2}(m_1 + m_2)\mathcal{M}_{13} - \frac{1}{2}m_1 \mathcal{M}_{14} + \frac{1}{2}(s + m_1 m_2)\mathcal{M}_{15},$$

$$\begin{aligned}
B_3 =& \frac{1}{2}\mathcal{M}_2 + \frac{1}{4}\mathcal{M}_3 + \frac{1}{4}\mathcal{M}_4 + \frac{1}{2}\mathcal{M}_5 + \frac{1}{2}\mathcal{M}_6 \\
& - \frac{1}{2}m_2\mathcal{M}_7 - \frac{1}{2}(m_1+m_2)\mathcal{M}_8 - \frac{1}{2}m_1\mathcal{M}_9 \\
& - \frac{1}{4}(m_1+m_2)\mathcal{M}_{10} - \frac{1}{4}m_1\mathcal{M}_{11} + \frac{1}{4}(s+m_1m_2)\mathcal{M}_{12} \\
& - \frac{1}{4}(m_1+m_2)\mathcal{M}_{13} - \frac{1}{4}m_1\mathcal{M}_{14} + \frac{1}{4}(s+m_1m_2)\mathcal{M}_{15}, \\
B_4 =& \frac{1}{4}\mathcal{M}_3 - \frac{1}{4}\mathcal{M}_4 + \frac{1}{2}\mathcal{M}_5 + \frac{1}{2}\mathcal{M}_6 - \frac{1}{2}m_2\mathcal{M}_7 \\
& - \frac{1}{4}(m_1+m_2)\mathcal{M}_{10} - \frac{1}{4}m_1\mathcal{M}_{11} + \frac{1}{4}(s+m_1m_2)\mathcal{M}_{12} \\
& + \frac{1}{4}(m_1+m_2)\mathcal{M}_{13} + \frac{1}{4}m_1\mathcal{M}_{14} - \frac{1}{4}(s+m_1m_2)\mathcal{M}_{15}, \\
B_5 =& -\mathcal{M}_1 + (m_2-m_1)\mathcal{M}_5 - m_1\mathcal{M}_6 - (s-m_1m_2)\mathcal{M}_7, \\
B_6 =& -\mathcal{M}_{10} - \mathcal{M}_{11} + m_2\mathcal{M}_{12} - \mathcal{M}_{13} - \mathcal{M}_{14} + m_2\mathcal{M}_{15}, \\
B_7 =& -\frac{1}{2}\mathcal{M}_{10} - \frac{1}{2}\mathcal{M}_{11} + \frac{1}{2}m_2\mathcal{M}_{12} + \frac{1}{2}\mathcal{M}_{13} + \frac{1}{2}\mathcal{M}_{14} - \frac{1}{2}m_2\mathcal{M}_{15}, \\
B_8 =& -\mathcal{M}_8 - \mathcal{M}_9 - \frac{1}{2}\mathcal{M}_{10} - \frac{1}{2}\mathcal{M}_{11} + \frac{1}{2}m_2\mathcal{M}_{12} \\
& - \frac{1}{2}\mathcal{M}_{13} - \frac{1}{2}\mathcal{M}_{14} + \frac{1}{2}m_2\mathcal{M}_{15}.
\end{aligned} \qquad (J.2)$$

For a gauge invariant amplitude, the following relations for the B_i hold:

$$k^2 B_1 + 2(k\cdot P)B_2 + 2(k\cdot q)B_3 + 2k^2 B_4 = 0,$$
$$B_5 + (k\cdot P)B_6 + k^2 B_7 + (k\cdot q)B_8 = 0.$$

Given these relations, one can eliminate two of the B_i (conventionally, one takes B_3 and B_5) and rewrite the amplitude in a manifestly gauge invariant form:

$$T_{fi} = i\bar{u}_2 \sum_{i=1}^{6} A_i M_i u_1 \qquad (J.3)$$

with the operator structures

$$M_1 = \frac{1}{2}\gamma_5\gamma_\mu\gamma_\nu F^{\mu\nu}, \qquad M_2 = 2\gamma_5 P_\mu(q - \frac{1}{2}k)_\nu F^{\mu\nu},$$
$$M_3 = \gamma_5\gamma_\mu q_\nu F^{\mu\nu}, \qquad M_4 = 2\gamma_5\gamma_\mu P_\nu F^{\mu\nu} - (m_1+m_2)M_1,$$
$$M_5 = \gamma_5 k_\mu q_\nu F^{\mu\nu}, \qquad M_6 = \gamma_5 k_\mu \gamma_\nu F^{\mu\nu},$$

where $F^{\mu\nu} = \epsilon^\mu k^\nu - \epsilon^\nu k^\mu$. The particular form of M_4 has been chosen such that

$$i\bar{u}_2 M_4 u_1 \to 2\sqrt{m_1 m_2}\, \mathbf{q}\cdot(\mathbf{k}\times\boldsymbol{\epsilon})$$

in the nonrelativistic limit, where the baryon masses are large compared to the meson masses and three-momenta.

The A_i are related to the B_i via

$$A_1 = B_1 - \frac{1}{2}(m_1 + m_2)B_6 ,$$

$$A_2 = \frac{2}{M_\phi^2 - t}B_2 ,$$

$$A_3 = -B_8 ,$$

$$A_4 = -\frac{1}{2}B_6 ,$$

$$A_5 = \frac{2}{s + u - m_1^2 - m_2^2}\left(B_1 - \frac{s - u + m_2^2 - m_1^2}{2(M_\phi^2 - t)}B_2 + 2B_4 \right) ,$$

$$A_6 = B_7 .$$

Following the conventions of Chew, Goldberger, Low and Nambu (CGLN) [144], and Berends, Donnachie and Weaver [145], we rewrite this once more, making use of the standard representation of spinors and Gamma matrices. In terms of Pauli spinors and matrices, one obtains

$$\frac{1}{8\pi\sqrt{s}}i\bar{u}_2 \sum_{i=1}^{6} A_i M_i u_1 = \chi_2^\dagger \mathbf{F} \chi_1, \tag{J.4}$$

where the matrix $\mathbf{F}$ reads

$$\mathbf{F} = i\sigma \cdot \mathbf{b}\,\mathcal{F}_1 + \sigma \cdot \hat{\mathbf{q}}\,\sigma \cdot (\hat{\mathbf{k}} \times \mathbf{b})\mathcal{F}_2 + i\sigma \cdot \hat{\mathbf{k}}\,\hat{\mathbf{q}} \cdot \mathbf{b}\,\mathcal{F}_3$$
$$+ i\sigma \cdot \hat{\mathbf{q}}\,\hat{\mathbf{q}} \cdot \mathbf{b}\,\mathcal{F}_4 - i\sigma \cdot \hat{\mathbf{q}}\,b_0\mathcal{F}_7 - i\sigma \cdot \hat{\mathbf{k}}\,b_0\mathcal{F}_8 .$$

Here, the hat over the three-vectors of course means normalization to a unit vector, and the four-vector b_μ is defined as

$$b_\mu = \epsilon_\mu - \frac{\epsilon \cdot \hat{\mathbf{k}}}{|\mathbf{k}|}k_\mu .$$

By substituting the standard representation of the Dirac spinors and matrices on the left-hand side of eq. (J.4), one finds the following expressions for the so-called CGLN-amplitudes $\mathcal{F}_i$:

$$\mathcal{F}_1 = (\sqrt{s} - m_1)\frac{N_1 N_2}{8\pi\sqrt{s}}\left[A_1 + \frac{k \cdot q}{\sqrt{s} - m_1}A_3 + (\sqrt{s} - m_2 - \frac{k \cdot q}{\sqrt{s} - m_1})A_4 - \frac{k^2}{\sqrt{s} - m_1}A_6 \right],$$

$$\mathcal{F}_2 = (\sqrt{s} + m_1)\frac{N_1 N_2}{8\pi\sqrt{s}}\frac{|\mathbf{q}||\mathbf{k}|}{(E_1 + m_1)(E_2 + m_2)}\left[-A_1 + \frac{k \cdot q}{\sqrt{s} + m_1}A_3 + (\sqrt{s} + m_2 - \frac{k \cdot q}{\sqrt{s} + m_1})A_4 \right.$$
$$\left. - \frac{k^2}{\sqrt{s} + m_1}A_6 \right],$$

$$\mathcal{F}_3 = (\sqrt{s} + m_1)\frac{N_1 N_2}{8\pi\sqrt{s}}\frac{|\mathbf{q}||\mathbf{k}|}{E_1 + m_1}\left[\frac{m_1^2 - s + \frac{1}{2}k^2}{\sqrt{s} + m_1}A_2 + A_3 - A_4 - \frac{k^2}{\sqrt{s} + m_1}A_5 \right],$$

$$\mathcal{F}_4 = (\sqrt{s} - m_1)\frac{N_1 N_2}{8\pi\sqrt{s}}\frac{|\mathbf{q}|^2}{E_2 + m_2}\left[\frac{s - m_1^2 - \frac{1}{2}k^2}{\sqrt{s} - m_1}A_2 + A_3 - A_4 + \frac{k^2}{\sqrt{s} - m_1}A_5 \right],$$

$$\mathcal{F}_7 = \frac{N_1 N_2}{8\pi\sqrt{s}}\frac{|\mathbf{q}|}{E_2 + m_2}\left[(m_1 - E_1)A_1 - \left(\frac{|\mathbf{k}|^2}{2k_0}(2k_0\sqrt{s} - 3k \cdot q) - \frac{\mathbf{q} \cdot \mathbf{k}}{2k_0}(2s - 2m_1^2 - k^2) \right)A_2 \right.$$
$$+ (q_0(\sqrt{s} - m_1) - k \cdot q)A_3 + (k \cdot q - q_0(\sqrt{s} - m_1) + (E_1 - m_1)(\sqrt{s} + m_2))A_4$$
$$\left. + (q_0 k^2 - k_0 k \cdot q)A_5 - (E_1 - m_1)(\sqrt{s} + m_1)A_6 \right],$$

$$\mathcal{F}_8 = \frac{N_1 N_2}{8\pi\sqrt{s}} \frac{|\mathbf{k}|}{E_1 + m_1} \bigg[(E_1 + m_1) A_1 + \left(\frac{|\mathbf{k}|^2}{2k_0}(2k_0\sqrt{s} - 3k\cdot q) - \frac{\mathbf{q}\cdot\mathbf{k}}{2k_0}(2s - 2m_1^2 - k^2) \right) A_2$$
$$+ \big(q_0(\sqrt{s} + m_1) - k\cdot q \big) A_3 + \big(k\cdot q - q_0(\sqrt{s} + m_1) + (E_1 + m_1)(\sqrt{s} - m_2) \big) A_4$$
$$- (q_0 k^2 - k_0 k\cdot q) A_5 - (E_1 + m_1)(\sqrt{s} - m_1) A_6 \bigg],$$

where

$$N_i = \sqrt{E_i + m_i}, \qquad E_i = \sqrt{\mathbf{p}_i^2 + m_i^2}, \qquad i = 1, 2.$$

We remark that, starting from eq. (J.3), we have utilized the Lorentz condition $k\cdot\epsilon = 0$. This is also valid when electroproduction is considered, since in that case the object ϵ^μ is proportional to the electron-photon vertex of QED, and the condition then follows from current conservation. Restricting ourselves to s-and p-waves, we can use the CGLN-amplitudes to arrive at the multipoles E_{0+}, M_{1+}, etc.:

$$\begin{pmatrix} E_{0+} \\ M_{1+} \\ M_{1-} \\ E_{1+} \end{pmatrix} = \int_{-1}^{1} dz \begin{pmatrix} \frac{1}{2}P_0 & -\frac{1}{2}P_1 & 0 & \frac{1}{6}P_{0,2} \\ \frac{1}{4}P_1 & -\frac{1}{4}P_2 & -\frac{1}{12}P_{0,2} & 0 \\ -\frac{1}{2}P_1 & \frac{1}{2}P_0 & \frac{1}{6}P_{0,2} & 0 \\ \frac{1}{4}P_1 & -\frac{1}{4}P_2 & \frac{1}{12}P_{0,2} & \frac{1}{10}P_{1,3} \end{pmatrix} \begin{pmatrix} \mathcal{F}_1 \\ \mathcal{F}_2 \\ \mathcal{F}_3 \\ \mathcal{F}_4 \end{pmatrix}$$

and

$$\begin{pmatrix} L_{0+} \\ L_{1+} \\ L_{1-} \end{pmatrix} = \frac{k_0}{|\mathbf{k}|} \int_{-1}^{1} dz \begin{pmatrix} \frac{1}{2}P_1 & \frac{1}{2}P_0 \\ \frac{1}{4}P_2 & \frac{1}{4}P_1 \\ \frac{1}{2}P_0 & \frac{1}{2}P_1 \end{pmatrix} \begin{pmatrix} \mathcal{F}_7 \\ \mathcal{F}_8 \end{pmatrix},$$

where $P_\ell \equiv P_\ell(z)$ are the usual Legendre polynomials and z is $\hat{\mathbf{q}}\cdot\hat{\mathbf{k}}$, i.e. the cosine of the scattering angle in the c.m. frame. Furthermore, the abbreviations

$$P_{0,2} = P_0 - P_2 \qquad \text{and} \qquad P_{1,3} = P_1 - P_3$$

were used.

A word on units: Since we use

$$\hbar = c = 1, \qquad e^2 = 4\pi\alpha$$

and normalize our Dirac spinors like $\bar{u}u = 2m$, the invariant amplitude B_5 has dimension GeV^{-1}, as can be seen, e.g., from the contribution of the graph corresponding to the 'Kroll-Ruderman' term. Therefore, the scattering amplitude T_{fi} is dimensionless (see eq. (J.1)), while the CGLN-amplitudes as well as the multipoles have dimension GeV^{-1}.

The unpolarized differential cross section for $\gamma p \to BM$ is given in terms of the CGLN-amplitudes as [145]

$$\frac{d\sigma}{d\Omega} = \frac{|\mathbf{q}|}{|\mathbf{k}|} \bigg\{ |\mathcal{F}_1|^2 + |\mathcal{F}_2|^2 + \frac{1}{2}|\mathcal{F}_3|^2 + \frac{1}{2}|\mathcal{F}_4|^2 + \mathrm{Re}(\mathcal{F}_1\mathcal{F}_4^*) + \mathrm{Re}(\mathcal{F}_2\mathcal{F}_3^*)$$
$$+ \big(\mathrm{Re}(\mathcal{F}_3\mathcal{F}_4^*) - 2\,\mathrm{Re}(\mathcal{F}_1\mathcal{F}_2^*) \big) \cos\theta - \left(\frac{1}{2}|\mathcal{F}_3|^2 + \frac{1}{2}|\mathcal{F}_4|^2 + \mathrm{Re}(\mathcal{F}_1\mathcal{F}_4^* + \mathcal{F}_2\mathcal{F}_3^*) \right) \cos^2\theta$$
$$- \mathrm{Re}(\mathcal{F}_3\mathcal{F}_4^*) \cos^3\theta \bigg\}. \qquad (\mathrm{J.5})$$

Acknowledgements

First of all, I would like to thank my Ph.D. supervisor Prof. Ulf-G. Meißner for giving me the possibility to work on this interesting subject of low-energy QCD, for his interest in my work, support and motivation. I also thank PD Dr. Akaki Rusetsky for proofreading and assessing this thesis.
Moreover, I want to express my gratitude to my long-time office-mates for invaluable help on countless occasions, many interesting discussions on physics, philosophy or nonsense, and for a pleasant and stimulating atmosphere in room 311: The first generation, Bugra Borasoy and Robin Nißler, who unfortunately both left physics in order to earn money, and the next generation, Maxim Mai and Christian Eilhardt, who will presumably both enjoy a great scientific career.
Of course, I am also grateful to everybody else at the institute, and to my friends and family. In particular, I would like to thank my parents, my brother Heiko, Ulrich Brüning and Sven Paulus for their repeated hospitality.

This research is part of the EU Integrated Infrastructure Initiative Hadron Physics Project under contract number RII3-CT-2004-506078. Work supported in part by DFG (SFB/TR 16, "Subnuclear Structure of Matter") and by the Helmholtz Association through funds provided to the virtual institute "Spin and strong QCD" (VH-VI-231).

Bibliography

[1] Th. W. Adorno and M. Horkheimer, "Dialektik der Aufklärung", S.Fischer Verlag (1988).

[2] S. Weinberg, arXiv:hep-th/9702027 (http://arxiv.org).

[3] S. Weinberg, "The Quantum Theory of Fields. Vol. 1: Foundations", Cambridge Univ. Pr. (1995).

[4] D. J. Gross and F. Wilczek, Phys. Rev. Lett. 30 (1973) 1343.

[5] H. D. Politzer, Phys. Rev. Lett. 30 (1973) 1346.

[6] I. Montvay and G. Münster, "Quantum Fields on a Lattice", Cambridge Univ. Pr. (1994).

[7] R. Gupta, arXiv:hep-lat/9807028.

[8] S. Weinberg, Physica A 96, (1979) 327.

[9] G. 't Hooft, Phys. Rev. D 14 (1976) 3432 [Erratum-ibid. D 18 (1978) 2199].

[10] E. Witten, Nucl. Phys. B 156 (1979) 269.

[11] K. Fujikawa, Phys. Rev. Lett. 42 (1979) 1195.

[12] C. Vafa and E. Witten, Nucl. Phys. B 234 (1984) 173.

[13] Y. Nambu, Phys. Rev. Lett. 4 (1960) 380.

[14] J. Goldstone, Nuovo Cim. 19 (1961) 154; J. Goldstone, A. Salam and S. Weinberg, Phys. Rev. 127 (1962) 965.

[15] Particle Data Group, http://pdg.lbl.gov.

[16] S. Weinberg, "The Quantum Theory of Fields. Vol. 2: Modern applications", Cambridge Univ. Pr. (1996).

[17] L. H. Ryder, "Quantum Field Theory", Cambridge Univ. Pr. (1985).

[18] M. E. Peskin and D. V. Schroeder, "An Introduction to Quantum Field Theory", Westview Press (1995).

[19] H. Leutwyler, Annals Phys. 235 (1994) 165 [arXiv:hep-ph/9311274].

[20] J. Gasser and H. Leutwyler, Ann. Phys. (New York) 158 (1984) 142.

[21] J. Gasser and H. Leutwyler, Nucl. Phys. B 250 (1985) 465.

[22] M. Gell-Mann, R. J. Oakes and B. Renner, Phys. Rev. 175 (1968) 2195.

[23] N. H. Fuchs, H. Sazdjian and J. Stern, Phys. Lett. B 269 (1991) 183; J. Stern, H. Sazdjian and N. H. Fuchs, Phys. Rev. D 47 (1993) 3814 [arXiv:hep-ph/9301244].

[24] S. L. Adler and R. F. Dashen, "Current Algebras and applications to particle physics", Benjamin, N.Y. (1968).

[25] J. Bijnens, G. Colangelo and G. Ecker, JHEP 9902 (1999) 020 [arXiv:hep-ph/9902437].

[26] S. L. Adler, Phys. Rev. 177 (1969) 2426.

[27] J. S. Bell and R. Jackiw, Nuovo Cim. A 60 (1969) 47.

[28] J. Wess and B. Zumino, Phys. Lett. B 37 (1971) 95.

[29] E. Witten, Nucl. Phys. B 223 (1983) 422.

[30] J. C. Collins, "Renormalization. An Introduction to Renormalization, the Renormalization Group, and the Operator Product Expansion", Cambridge Univ. Pr. (1984).

[31] J. Gasser, M. E. Sainio and A. Svarc, Nucl. Phys. B 307, (1988) 779.

[32] A. Krause, Helv. Phys. Acta 63 (1990) 3.

[33] F. E. Close and R. G. Roberts, Phys. Lett. B 316 (1993) 165 [arXiv:hep-ph/9306289]; B. Borasoy, Phys. Rev. D 59 (1999) 054021 [arXiv:hep-ph/9811411].

[34] E. E. Jenkins and A. V. Manohar, Phys. Lett. B 255 (1991) 558.

[35] V. Bernard, N. Kaiser, J. Kambor and U.-G. Meißner, Nucl. Phys. B 388 (1992) 315.

[36] V. Bernard, N. Kaiser and U.-G. Meißner, Int. J. Mod. Phys. E 4 (1995) 193 [arXiv:hep-ph/9501384].

[37] V. Bernard, Prog. Part. Nucl. Phys. 60 (2008) 82 [arXiv:0706.0312 [hep-ph]].

[38] T. Becher and H. Leutwyler, Eur. Phys. J. C 9 (1999) 643 [hep-ph/9901384].

[39] H. B. Tang, [hep-ph/9607436].

[40] P. J. Ellis and H. B. Tang, Phys. Rev. C 57 (1998) 3356 [hep-ph/9709354].

[41] B. Kubis and U.-G. Meißner, Nucl. Phys. A 679 (2001) 698 [hep-ph/0007056].

[42] G. Ecker, J. Gasser, A. Pich and E. de Rafael, Nucl. Phys. B 321 (1989) 311.

[43] V. Bernard, N. Kaiser and U.-G. Meißner, Nucl. Phys. A 615 (1997) 483 [hep-ph/9611253].

[44] E. E. Jenkins and A. V. Manohar, Phys. Lett. B 259 (1991) 353.

[45] T. R. Hemmert, B. R. Holstein and J. Kambor, J. Phys. G 24 (1998) 1831 [arXiv:hep-ph/9712496].

[46] V. Pascalutsa and M. Vanderhaeghen, Phys. Rev. D 73 (2006) 034003 [arXiv:hep-ph/0512244]; V. Pascalutsa, Prog. Part. Nucl. Phys. 61 (2008) 27 [arXiv:0712.3919 [nucl-th]].

[47] V. Bernard, T. R. Hemmert and U.-G. Meißner, Phys. Lett. B 565 (2003) 137 [arXiv:hep-ph/0303198].

[48] V. Bernard, T. R. Hemmert and U.-G. Meißner, Phys. Lett. B 622 (2005) 141 [arXiv:hep-lat/0503022].

[49] C. Hacker, N. Wies, J. Gegelia and S. Scherer, Phys. Rev. C 72 (2005) 055203 [arXiv:hep-ph/0505043].

[50] G. Ecker, J. Gasser, H. Leutwyler, A. Pich and E. de Rafael, Phys. Lett. B 223 (1989) 425.

[51] P. C. Bruns and U.-G. Meißner, Eur. Phys. J. C 40 (2005) 97 [hep-ph/0411223].

[52] B. Borasoy and U.-G. Meißner, Int. J. Mod. Phys. A 11 (1996) 5183 [hep-ph/9511320].

[53] R. J. Eden, P. V. Landshoff, D. I. Olive and J. C. Polkinghorne, "The Analytic S-Matrix", Cambridge Univ. Pr. (1966).

[54] R. Nißler, Ph.D. thesis, University of Bonn (2008).

[55] J. A. Oller and E. Oset, Nucl. Phys. A 620 (1997) 438 [Erratum-ibid. A 652 (1999) 407] [arXiv:hep-ph/9702314]; J. A. Oller, E. Oset and J. R. Pelaez, Phys. Rev. D 59 (1999) 074001 [Erratum-ibid. D 60 (1999) 099906] [arXiv:hep-ph/9804209].

[56] N. Kaiser, P. B. Siegel and W. Weise, Phys. Lett. B 362 (1995) 23 [arXiv:nucl-th/9507036]; N. Kaiser, P. B. Siegel and W. Weise, Nucl. Phys. A 594 (1995) 325 [arXiv:nucl-th/9505043].

[57] J. A. Oller and U.-G. Meißner, Phys. Lett. B 500 (2001) 263 [arXiv:hep-ph/0011146]; U.-G. Meißner and J. A. Oller, Nucl. Phys. A 673 (2000) 311 [arXiv:nucl-th/9912026].

[58] E. E. Kolomeitsev and M. F. M. Lutz, Phys. Lett. B 585 (2004) 243 [arXiv:nucl-th/0305101].

[59] D. Jido, J. A. Oller, E. Oset, A. Ramos and U.-G. Meißner, Nucl. Phys. A 725 (2003) 181 [arXiv:nucl-th/0303062].

[60] J. Nieves and E. Ruiz Arriola, Nucl. Phys. A 679 (2000) 57 [arXiv:hep-ph/9907469].

[61] U.-G. Meißner, V. Mull, J. Speth and J. W. van Orden, Phys. Lett. B 408 (1997) 381 [hep-ph/9701296].

[62] H. W. Hammer and M. J. Ramsey-Musolf, Phys. Lett. B 416 (1998) 5 [hep-ph/9703406].

[63] T. Fuchs, M. R. Schindler, J. Gegelia and S. Scherer, Phys. Lett. B 575 (2003) 11 [hep-ph/0308006]

[64] J. Schweizer, Diploma thesis, University of Bern, 2000.

[65] N. Fettes, U.-G. Meißner and S. Steininger, Nucl. Phys. A 640 (1998) 199 [hep-ph/9803266]

[66] K. Kawarabayashi and M. Suzuki, Phys. Rev. Lett. 16 (1966) 255; Fayyazuddin and Riazuddin, Phys. Rev. 147 (1966) 1071.

[67] D. Djukanovic, M. R. Schindler, J. Gegelia, G. Japaridze and S. Scherer, Phys. Rev. Lett. 93 (2004) 122002 [hep-ph/0407239].

[68] M. R. Schindler, T. Fuchs, J. Gegelia and S. Scherer, Phys. Rev. C 75 (2007) 025202 [nucl-th/0611083].

[69] M. Gari and U. Kaulfuss, Phys. Lett. B 138 (1984) 29.

[70] D. Lehmann and G. Prezeau, Phys. Rev. D 65 (2002) 016001 [hep-ph/0102161].

[71] S. Sasaki, T. Blum and S. Ohta, Phys. Rev. D 65 (2002) 074503 [arXiv:hep-lat/0102010].

[72] N. Mathur et al., Phys. Lett. B 605 (2005) 137 [arXiv:hep-ph/0306199].

[73] D. B. Leinweber, W. Melnitchouk, D. G. Richards, A. G. Williams and J. M. Zanotti, Lect. Notes Phys. 663 (2005) 71 [arXiv:nucl-th/0406032].

[74] D. Guadagnoli, M. Papinutto and S. Simula, Phys. Lett. B 604 (2004) 74 [arXiv:hep-lat/0409011].

[75] K. Sasaki and S. Sasaki, Phys. Rev. D 72 (2005) 034502 [arXiv:hep-lat/0503026].

[76] K. Sasaki, S. Sasaki and T. Hatsuda, Phys. Lett. B 623 (2005) 208 [arXiv:hep-lat/0504020].

[77] T. Burch, C. Gattringer, L. Y. Glozman, C. Hagen, D. Hierl, C. B. Lang and A. Schäfer, Phys. Rev. D 74 (2006) 014504 [arXiv:hep-lat/0604019].

[78] N. Fettes, U.-G. Meißner, M. Mojžiš and S. Steininger, Annals Phys. 283 (2000) 273 [Erratum-ibid. 288 (2001) 249] [arXiv:hep-ph/0001308].

[79] B. Borasoy and U.-G. Meißner, Annals Phys. 254 (1997) 192 [arXiv:hep-ph/9607432].

[80] J. Gasser and A. Zepeda, Nucl. Phys. B 174 (1980) 445.

[81] O. Krehl, C. Hanhart, S. Krewald and J. Speth, Phys. Rev. C 62 (2000) 025207 [arXiv:nucl-th/9911080].

[82] V. Bernard, N. Kaiser and U.-G. Meißner, Nucl. Phys. B 457 (1995) 147 [arXiv:hep-ph/9507418].

[83] E. Hernandez, E. Oset and M. J. Vicente Vacas, Phys. Rev. C 66 (2002) 065201 [arXiv:nucl-th/0209009].

[84] A. Holl, P. Maris, C. D. Roberts and S. V. Wright, arXiv:nucl-th/0512048.

[85] I. P. Cavalcante, M. R. Robilotta, J. Sa Borges, D. de O.Santos and G. R. S. Zarnauskas, Phys. Rev. C 72 (2005) 065207 [arXiv:hep-ph/0507147].

[86] G. Colangelo and S. Dürr, Eur. Phys. J. C 33 (2004) 543 [arXiv:hep-lat/0311023].

[87] V. Bernard, T. R. Hemmert and U.-G. Meißner, Nucl. Phys. A 732 (2004) 149 [arXiv:hep-ph/0307115].

[88] M. Procura, B. U. Musch, T. Wollenweber, T. R. Hemmert and W. Weise, Phys. Rev. D 73 (2006) 114510 [arXiv:hep-lat/0603001].

[89] N. Kaiser, T. Waas and W. Weise, Nucl. Phys. A 612 (1997) 297 [arXiv:hep-ph/9607459]; J. Caro Ramon, N. Kaiser, S. Wetzel and W. Weise, Nucl. Phys. A 672 (2000) 249 [arXiv:nucl-th/9912053].

[90] S. D. Bass, S. Wetzel and W. Weise, Nucl. Phys. A 686 (2001) 429 [arXiv:hep-ph/0007293].

[91] B. Borasoy, E. Marco and S. Wetzel, Phys. Rev. C 66 (2002) 055208 [arXiv:hep-ph/0212256].

[92] F. Gross and D. O. Riska, Phys. Rev. C 36 (1987) 1928.

[93] Y. Surya and F. Gross, Phys. Rev. C 53 (1996) 2422.

[94] A. N. Kvinikhidze and B. Blankleider, Phys. Rev. C 60 (1999) 044003 [arXiv:nucl-th/9901001].

[95] A. N. Kvinikhidze and B. Blankleider, Phys. Rev. C 60 (1999) 044004 [arXiv:nucl-th/9901002].

[96] C. H. M. van Antwerpen and I. R. Afnan, Phys. Rev. C 52 (1995) 554 [arXiv:nucl-th/9407038].

[97] H. Haberzettl, Phys. Rev. C 56 (1997) 2041 [arXiv:nucl-th/9704057].

[98] J. A. Oller, Phys. Lett. B 426 (1998) 7 [arXiv:hep-ph/9803214].

[99] E. Marco, S. Hirenzaki, E. Oset and H. Toki, Phys. Lett. B 470 (1999) 20 [arXiv:hep-ph/9903217].

[100] B. Borasoy and R. Nißler, Eur. Phys. J. A 19 (2004) 367 [arXiv:hep-ph/0309011].

[101] B. Borasoy and R. Nißler, Nucl. Phys. A 740 (2004) 362 [arXiv:hep-ph/0405039].

[102] H. Haberzettl, K. Nakayama and S. Krewald, Phys. Rev. C 74 (2006) 045202 [arXiv:nucl-th/0605059].

[103] R. E. Cutkosky, Phys. Rev. 96 (1954) 1135.

[104] J. A. Oller, E. Oset and A. Ramos, Prog. Part. Nucl. Phys. 45 (2000) 157 [arXiv:hep-ph/0002193].

[105] T. Sekihara, T. Hyodo and D. Jido, Phys. Lett. B 669 (2008) 133 [arXiv:0803.4068 [nucl-th]].

[106] S. Capstick and N. Isgur, Phys. Rev. D 34 (1986) 2809.

[107] L. Y. Glozman, W. Plessas, K. Varga and R. F. Wagenbrunn, Phys. Rev. D 58 (1998) 094030 [arXiv:hep-ph/9706507].

[108] U. Löring, K. Kretzschmar, B. C. Metsch and H. R. Petry, Eur. Phys. J. A 10 (2001) 309 [arXiv:hep-ph/0103287].

[109] K. H. Glander et al., Eur. Phys. J. A 19 (2004) 251 [arXiv:nucl-ex/0308025].

[110] R. Lawall et al., Eur. Phys. J. A 24 (2005) 275 [arXiv:nucl-ex/0504014].

[111] J. W. C. McNabb et al. [CLAS Collaboration], Phys. Rev. C 69 (2004) 042201 [arXiv:nucl-ex/0305028].

[112] R. Bradford et al. [CLAS Collaboration], Phys. Rev. C 73 (2006) 035202 [arXiv:nucl-ex/0509033].

[113] R. Castelijns et al. [CBELSA/TAPS Collaboration], arXiv:nucl-ex/0702033.

[114] R. G. T. Zegers et al. [LEPS Collaboration], Phys. Rev. Lett. 91 (2003) 092001 [arXiv:nucl-ex/0302005].

[115] A. V. Anisovich, A. Sarantsev, O. Bartholomy, E. Klempt, V. A. Nikonov and U. Thoma, Eur. Phys. J. A 25 (2005) 427 [arXiv:hep-ex/0506010].

[116] A. V. Sarantsev, V. A. Nikonov, A. V. Anisovich, E. Klempt and U. Thoma, Eur. Phys. J. A 25 (2005) 441 [arXiv:hep-ex/0506011].

[117] A. Usov and O. Scholten, Phys. Rev. C 72 (2005) 025205 [arXiv:nucl-th/0503013].

[118] C. E. Carlson and J. L. Poor, Phys. Rev. D 38 (1988) 2758.

[119] D. B. Leinweber, T. Draper and R. M. Woloshyn, Phys. Rev. D 46 (1992) 3067 [arXiv:hep-lat/9208025].

[120] E. Oset and A. Ramos, Nucl. Phys. A 635 (1998) 99 [arXiv:nucl-th/9711022].

[121] M. F. M. Lutz and E. E. Kolomeitsev, Nucl. Phys. A 700 (2002) 193 [arXiv:nucl-th/0105042].

[122] J. C. Nacher, E. Oset, H. Toki and A. Ramos, Phys. Lett. B 461 (1999) 299 [arXiv:nucl-th/9902071].

[123] J. Caro Ramon, N. Kaiser, S. Wetzel and W. Weise, Nucl. Phys. A 672 (2000) 249 [arXiv:nucl-th/9912053].

[124] J. C. Nacher, E. Oset, H. Toki and A. Ramos, Phys. Lett. B 455 (1999) 55 [arXiv:nucl-th/9812055].

[125] J. A. Oller and E. Oset, Nucl. Phys. A 620 (1997) 438 [Erratum-ibid. A 652 (1999) 407] [arXiv:hep-ph/9702314].

[126] C. Bennhold, T. Mart and D. Kusno, arXiv:nucl-th/9703004.

[127] B. Ananthanarayan, G. Colangelo, J. Gasser and H. Leutwyler, Phys. Rept. 353 (2001) 207 [arXiv:hep-ph/0005297].

[128] P. Buettiker, S. Descotes-Genon and B. Moussallam, Eur. Phys. J. C 33 (2004) 409 [arXiv:hep-ph/0310283].

[129] T. Becher and H. Leutwyler, JHEP 0106 (2001) 017 [arXiv:hep-ph/0103263].

[130] T. M. Knasel et al., Phys. Rev. D 11 (1975) 1.

[131] R. D. Baker et al., Nucl. Phys. B 141 (1978) 29.

[132] B. Borasoy, U.-G. Meißner and R. Nißler, Phys. Rev. C 74 (2006) 055201 [arXiv:hep-ph/0606108].

[133] R. D. Baker et al., Nucl. Phys. B 145 (1978) 402.

[134] A. Baldini, V. Flaminio, W. G. Moorhead, and D. R. O. Morrison, in: Landolt-Börnstein, Vol. 12a, ed. H. Schopper (Springer, Berlin, 1988).

[135] A. V. Anisovich, V. Kleber, E. Klempt, V. A. Nikonov, A. V. Sarantsev and U. Thoma, arXiv:0707.3596 [hep-ph].

[136] T. Inoue, E. Oset and M. J. Vicente Vacas, Phys. Rev. C 65 (2002) 035204 [arXiv:hep-ph/0110333].

[137] A. Motzke, C. Elster and C. Hanhart, Phys. Rev. C 66 (2002) 054002 [arXiv:nucl-th/0207047].

[138] M. Frink and U.-G. Meißner, JHEP 0407 (2004) 028 [arXiv:hep-lat/0404018].

[139] J. A. Oller, Eur. Phys. J. A 28 (2006) 63 [arXiv:hep-ph/0603134].

[140] N. Beisert, Diploma thesis, TU München (2001).

[141] D. Djukanovic, J. Gegelia and S. Scherer, Eur. Phys. J. A 29 (2006) 337 [arXiv:hep-ph/0604164].

[142] G. 't Hooft and M. J. G. Veltman, Nucl. Phys. B 153 (1979) 365.

[143] R. Mertig, M. Bohm and A. Denner, Comput. Phys. Commun. 64 (1991) 345. http://www.feyncalc.org

[144] G. F. Chew, M. L. Goldberger, F. E. Low and Y. Nambu, Phys. Rev. 106 (1957) 1345.

[145] F. A. Berends, A. Donnachie and D. L. Weaver, Nucl. Phys. B 4 (1967) 1.

I want morebooks!

Buy your books fast and straightforward online - at one of the world's fastest growing online book stores! Environmentally sound due to Print-on-Demand technologies.

Buy your books online at
www.get-morebooks.com

Kaufen Sie Ihre Bücher schnell und unkompliziert online – auf einer der am schnellsten wachsenden Buchhandelsplattformen weltweit!
Dank Print-On-Demand umwelt- und ressourcenschonend produziert.

Bücher schneller online kaufen
www.morebooks.de

OmniScriptum Marketing DEU GmbH
Heinrich-Böcking-Str. 6-8
D - 66121 Saarbrücken
Telefax: +49 681 93 81 567-9

info@omniscriptum.com
www.omniscriptum.com

Printed by Books on Demand GmbH, Norderstedt / Germany